Environment and Politics

Third edition

Timothy Doyle and Doug McEachern

LONDON AND NEW YORK

First published 1998
Reprinted 1999
Second edition 2001
Third edition 2008
by Routledge
2 Park Square, Milton Park, Abingdon, Oxon, OX14 4RN

Simultaneously published in the USA and Canada
by Routledge
270 Madison Avenue, New York, NY 10016

Routledge is an imprint of the Taylor & Francis Group

Typeset in Times New Roman by
Keystroke, 28 High Street, Tettenhall, Wolverhampton
Printed and bound in Great Britain by
The Cromwell Press, Trowbridge, Wiltshire

British Library Cataloguing in Publication Data
A catalogue record for this book is available from the British Library

Library of Congress Cataloging in Publication Data
Doyle, Timothy, 1960–
Environment and politics / Timothy Doyle and Doug McEachern. – 3rd ed.
p. cm.
Includes bibliographical references and index.
ISBN 978-0-415-38052-2 (hardcover) – ISBN 978-0-415-38051-5 (softcover).
1. Environmental policy. 2. Green movement–Political aspects.
I. McEachern, Doug. II. Title.
GE170.D69 2007
363.7–dc22
2007000043

ISBN10: 0–415–38051–0 (pbk)
ISBN10: 0–415–38052–9 (hbk)
ISBN13: 978–0–415–38051–5 (pbk)
ISBN13: 978–0–415–38052–2 (hbk)

Environment and Politics

Third edition

Environment and Politics is a concise introduction to the study of environmental politics, explaining the key concepts, conflicts, political systems and the practices of policy making to solve environmental problems worldwide. The authors examine the institutional responses of parliaments, administrative, legal and electoral systems; the more informal politics of non-governmental organisations and social movements; and the politics of business as it responds to the greening of our society.

This third edition provides a large amount of new material and expands the discussion of differences in environmental politics between liberal democracies, military dictatorships and one-party states. It looks at future developments for effective local and international environmental diplomacy and both global and region-specific problem solving. A selection of case studies investigating theories and practice of environmental politics in countries such as China, Thailand, Burma, Iran, Nigeria, Cuba, The Philippines, India, Ecuador, Mexico, Hungary and Bosnia–Herzegovina are introduced, and the book importantly addresses the growth in green politics outside of mainstream, liberal-democratic, Western models of politics. Also added are case studies on issues such as genetically modified organisms (GMOs); food sovereignty, road-building, environmental and energy security, alternative technology, mining, water and dam-building, nuclear power and disarmament. The growing nexus between religious NGOs and the green movement, in Christian, Muslim and Buddhist contexts is discussed and recent developments in the new social movement theory are introduced, with philosophical and political differences articulated between three forms of environmentalism: post-materialism; post-industrialism; and post-colonialism. In many ways *the* defining feature of this third edition is extended research into business responses to the environment in relation to the powerful current issue of climate change, and the impact of globalisation on all forms of green politics.

Environment and Politics – as with its first two editions – continues to draw on a wealth of original examples – from North and South America, Europe, Asia, Africa and Australia – and provides the reader with a greater understanding of international, national and local environmental politics. It also looks at future developments for effective local and international environmental diplomacy and both global and region-specific problem solving.

Timothy Doyle is Professor of Politics and International Relations, SPIRE, Keele University, United Kingdom; **Doug McEachern** is Professor, and Pro-Vice Chancellor (Research) at the University of Western Australia.

Routledge Introductions to Environment Series

Published and Forthcoming Titles

Titles under Series Editors:
Rita Gardner and A.M. Mannion

Environmental Science texts

Atmospheric Processes and Systems
Natural Environmental Change
Biodiversity and Conservation
Ecosystems
Environmental Biology
Using Statistics to Understand The Environment
Coastal Systems
Environmental Physics
Environmental Chemistry
Biodiversity and conservation, 2nd Edition
Ecosystems, 2nd Edition

Titles under Series Editor:
David Pepper

Environment and Society texts

Environment and Politics
Environment and Philosophy
Environment and Social Theory
Energy, Society and Environment, 2nd edition
Environment and Tourism
Gender and Environment
Environment and Business
Environment and Politics, 2nd edition
Environment and Law
Environment & Society
Environmental Policy
Representing the Environment
Sustainable Development
Environment and Social Theory, 2nd edition
Environmental Values
Environment and Politics, 3rd Edition

Contents

Series editor's preface *Environment and Society titles*

The modern environmentalist movement grew hugely in the last third of the twentieth century. It reflected popular and academic concerns about the local and global degradation of the physical environment which was increasingly being documented by scientists (and which is the subject of the companion series to this, Environmental Science). However it soon became clear that reversing such degradation was not merely a technical and managerial matter: merely knowing about environmental problems did not of itself guarantee that governments, businesses or individuals would do anything about them. It is now acknowledged that a critical understanding of socio-economic, political and cultural processes and structures is central in understanding environmental problems and establishing environmentally sustainable development. Hence the maturing of environmentalism has been marked by prolific scholarship in the social sciences and humanities, exploring the complexity of society-environment relationships.

Such scholarship has been reflected in a proliferation of associated courses at undergraduate level. Many are taught within the 'modular' or equivalent organisational frameworks which have been widely adopted in higher education. These frameworks offer the advantages of flexible undergraduate programmes, but they also mean that knowledge may become segmented, and student learning pathways may arrange knowledge segments in a variety of sequences – often reflecting the individual requirements and backgrounds of each student rather than more traditional discipline-bound ways of arranging learning.

The volumes in this Environment and Society series of textbooks mirror this higher educational context, increasingly encountered in the early twenty-first century. They provide short, topic-centred texts on social science and humanities subjects relevant to contemporary society-environment relations. Their content and approach reflect the fact that each will be read by students from various disciplinary backgrounds, taking in not only social sciences and humanities but others such as physical and natural sciences. Such a readership is not always familiar with the disciplinary background to a topic, neither are readers necessarily going on to further develop their interest in the topic. Additionally, they cannot all automatically be thought of as having reached a similar stage in their studies – they may be first-, second- or third-year students.

The authors and editors of this series are mainly established teachers in higher education. Finding that more traditional integrated environmental studies and specialised texts do not always meet their own students' requirements, they have often had to write course materials more appropriate to the needs of the flexible undergraduate programme. Many of the volumes in this series represent in modified form the fruits of such labours, which all students can now share.

Much of the integrity and distinctiveness of the Environment and Society titles derives from their characteristic approach. To achieve the right mix of flexibility, breadth and depth, each volume is designed to create maximum accessibility to readers from a variety of backgrounds and attainment. Each leads into its topic by giving some necessary basic grounding, and leaves it usually by pointing towards areas for further potential development and study. There is introduction to the real-world context of the text's main topic, and to the basic concepts and questions in social sciences/humanities which are most relevant. At the core of the text is some exploration of the main issues. Although limitations are imposed here by the need to retain a book length and format affordable to students, some care is taken to indicate how the themes and issues presented may become more complicated, and to refer to the cognate issues and concepts that would need to be explored to gain deeper understanding. Annotated reading lists, case studies, overview diagrams, summary charts and self-check questions and exercises are among the pedagogic devices which we try to encourage our authors to use, to maximise the 'student friendliness' of these books.

Hence we hope that these concise volumes provide sufficient depth to maintain the interest of students with relevant backgrounds. At the same time, we try to ensure that they sketch out basic concepts and map their territory in a stimulating and approachable way for students to whom the whole area is new. Hopefully, the list of Environment and Society titles will provide modular and other students with an unparalleled range of perspectives on society-environment problems: one which should also be useful to students at both postgraduate and pre-higher education levels.

David Pepper
May 2000

Series International Advisory Board

Plates

Figures and tables

Figures

Tables

Boxes

Preface to third edition

We awoke on October 4 in the Nigerian capital, Abuja, to yet more news of kidnappings: this time seven foreign oil-workers had been taken from their compound in Eket, in the oil-rich Niger Delta. Two Nigerian security guards were believed to have been killed in the attack. These kidnappings followed directly on the heels of another attack earlier in the week against staff of a Royal Dutch Shell contractor. On this occasion, 70 gunmen attacked Shell facilities such as fuel-carrying barges, as well as taking 25 staff hostage. Three soldiers were confirmed dead and 7 more were missing after the attack.

Accessing various foreign press outlets via the internet, we attempted to put together the limited pieces of information available to try to make some sense of what was going on in the Delta. Most articles listed the dead and the missing. Some argued that such disruptions had cut Africa's largest suppliers of crude oil by nearly a quarter over the past year causing several spikes in world oil prices. Others spoke of the struggle to control the lucrative oil smuggling business, or the attraction of criminals to the 'lure of ransoms'.

Few outlets, however, attempted to explain the underlying causes of these events. *The Washington Post* did concede that 'the action was taken in revenge for attacks by soldiers on local communities', whilst CNN's coverage was the most comprehensive in relative terms, arguing that 'violence in the Delta is rooted in poverty, corruption and lawlessness. Most inhabitants of the wetlands region, which is almost the size of England, have seen few benefits for five decades of oil extraction that has damaged their environment. Their resentment towards the oil industry breeds militancy.'

Of course, social and environmental problems in Nigeria have their origins beyond the last fifty years. The fight for oil is only the most recent manifestation. Sub-Saharan Africa was the main site for the global slave trade which lasted for over four hundred years up until the nineteenth century. African coastal dwellers were forcefully employed by the Europeans to penetrate further inland, capturing more slaves as they went. The ethno-religious divide in Nigeria today, in part, reflects the political economy of this early slave trade, with Christians predominant in the South, while the North is largely Muslim. With the advent of a vaccine for yellow fever (which paradoxically had protected Africans from the full extent of European colonialism) the 'Race for Africa' commenced in earnest in the late years of the nineteenth century. In Nigeria's case, it was Britain which was the coloniser, cobbling together a vast state out of disparate tribal and religious cultures, which would always prove problematic.

Despite the end of state-centred colonialism with the withdrawal of Britain in 1960, a new era of colonialism rooted in transnational industry (like the global slave trade) began in 1957: 'black gold' was discovered in a joint venture between Royal Dutch Shell and British Petroleum. Oil displaced people as Nigeria's most traded commodity; but, in many ways, the commercial culture of the slave trade remained, based on principles of exploitation, cruelty and short-term profiteering. After independence, the multi-ethnic nation-state of Nigeria has continued to exist, struggling between periods of military dictatorship and representative democracy. Today, oil accounts for 95 per cent of Nigeria's foreign income, and 85 per cent of total annual revenue. Due to the centrality of oil revenue to the Nigerian economy, the government works closely with transnational extraction companies, suppressing resistance to ensure smooth operations (Harkin 2006). Despite the exorbitant profits being made by the oil industry in Nigeria, 70 per cent of its 120 million people live below the poverty line, with most local people in the Delta living without medical and educational infrastructure (Mojavu 2003).

The story the plight of the Ogoni people of the Niger Delta (one of the groups of people indigenous to the area) reached the international spotlight in 1995, when Ken Saro-Wiwa and other members of the 'Ogoni Nine' were sentenced to death by the military dictatorship for their opposition to Shell's operations in the Delta. Before his trial and subsequent execution, he made the following speech:

> Today, the Ogoni people are involved in two grim wars. The first is the thirty-five year old ecological war waged by the multi-national oil companies, Shell and Chevron. In this sophisticated and

> unconventional war . . . the men, women and children die; flora, fauna and fish perish, the air and water are poisoned, and finally, the land dies. The second war is a political war of tyranny, oppression and greed designed to dispossess the Ogoni people of their rights and their wealth and subject them to abject poverty, slavery, dehumanisation and extinction.
>
> (Saro-Wiwa in Oskwa 2005: 16)

Many years after the death of Saro-Wiwa and the other members of the Ogoni Nine, the struggle for the people of the Niger Delta continues despite the military regime coming to an end. Although Nigeria's parliament has now been democratised, its environment has not. This fact explains the logo of Nigeria's pre-eminent environmental organisation, Environmental Rights Action (ERA, Friends of the Earth Nigeria). In its fight for 'democratising the environment', Nnimmo Bassey, the Executive Director of ERA, launched *The Shell Report* which documents continuing abuses in Nigeria. Bassey presents a case in which he lists thousands of Nigerians dying in military and extra-military conflicts associated with the oil industry every year. He writes:

> While ERA issues this report in commemoration of the 10th anniversary of the execution of the Ogoni patriots, we call the attention of all to the fact that the battle for a safe Niger Delta environment is far from being won. Obsolete pipes are still being used by the oil corporations; spills remain a matter of routine; social and moral disruptions are entrenched; toxic gases continue to be released through gas flares; poverty has turned even more virulent and life is still short and brutish. Moreover, the demands of the Ogoni in the Ogoni Bill of Rights and those of other ethnic nationalities remain scornfully unattended to.
>
> (Bassey in Oskwa 2005: 3)

Apart from building networks of solidarity in grassroots communities, much of ERA's work today is oriented around launching legal actions against transnational extractive companies both in Nigeria and in the countries of the firms' origins. In places on earth where it is difficult to fight for human and environmental justice, it is often necessary to 'internationalise' the cause in a bid to gain support. There have been few better exponents of this than ERA. In a bid to attract a transnational audience, often the 'stories' of the local people have to be repackaged and framed in a way which can be 'interpreted' by people who come from the more affluent world, who have more economic and political power to wield. In our meetings in Abuja, one particular moment exemplifies this point more than any other. At a meeting of Friends of the Earth International, an elder of the Ogoni people (name withheld) was asked if he saw the resistance of his people as part of a broader struggle to respond to global climate change. He replied that he had never heard of this concept, and sought clarification as to what the connections were. It was explained to him that climate change was being brought about by an over-reliance on fossil fuels. Climate change campaigners sought to replace these fuels with more alternative forms of energy. Ultimately, this would loosen the stranglehold on the global economy by oil extractive industries such as Shell. The old man stated that if this connection helped his people's cause in any way – by getting their message out to an international audience – then he was willing to endorse it; to take any possible pathway which would strengthen his people's claims.

From a position of powerlessness, we can understand the Ogoni Elder's willingness to 'reframe' his reality in a bid to get more support from affluent world environmental NGOs. But there may also be a loss when what is effectively a post-colonialist story of abject poverty is replaced with a post-industrialist one of the affluent world. A critical view of the predominant climate change discourse, for example, is that it takes much of the politics – the conflict – out of environmental resource issues, providing a polite filter between human action and human consequence; taking the direct and instrumental power relationships out of the equation. It is no longer people against people: the exploiters versus the exploited, or in this case, the polluters versus the polluted. Rather, although people are still the initiators, they are cast in a far more oblique light, often unwittingly setting off a calamitous, climactic punishment for all. A force of nature is, in the end, the nemesis, whereas the initiators, the environmental degraders, are in relative safety, at a convenient one-step removed from the atrocities inflicted upon the many. Also, by constructing the concept of an environmental 'day of judgment' for all, all humans (all creation) are cast equally as victims; not differentiating between the perpetrators and fatalities (Doyle and Doherty 2006).

These different types of modern environmentalisms are constantly engaged within this edition. There is no one environmentalism, nor is there one environmental movement which champions its principles in the world today. Environmental movements, groups, organisations, political parties, governments, institutions and corporations have responded to the politics of environmental concern with remarkable vigour and diversity. The politics of the environment, and the ideas which inform it, differs enormously in different parts of the globe, though there is also interplay between these different mind-sets as the world becomes increasingly dominated by the near ubiquitous forces of globalisation. These differences and tensions characterise this edition.

The cover of this third edition was chosen to illustrate the global and local dimensions of the most pressing environmental issues and problems. Whereas, in the affluent world, natural gas (usually a by-product of oil production) is utilised as an additional source of fuel to oil, in poorer countries like Nigeria it is often cheaper in the short term, though hardly environmentally sustainable, for the oil companies to burn it. Also, some of these gas flares are the result of militants who blow up oil installations in their attempts to disrupt the industry. As a consequence, all across the Niger Delta, people must live alongside these jets of burning gas, with some still burning for over twenty-five years. The two women on the front cover are drying tapioca near a continuously burning jet of gas in Warri, Nigeria. Although the jet of gas behind them symbolises the 'big politics' of global energy security, they must continue their everyday struggles to feed and clothe their families with the limited environmental 'goods' available to them.

Tim Doyle
Doug McEachern
Abuja, Nigeria
October 2006

Preface to second edition

Since the publication of the first edition of *Environment and Politics* in 1998 there have been many striking events to illustrate the new ways in which environmental politics are contested. The most important development from a Western, party political perspective was the election of the Greens to Federal Government in Germany, in partnership with the Socialist Democratic Party. What a dramatic transition this is when compared with the early days of green concern when environmentalists were regarded with deep suspicion and hostility by mainstream political forces.

There is no doubt that environmental concern has long evolved from the exclusively outsider politics of protest and conflict (although this still occurs frequently) to being incorporated into the platforms of political parties and the mission statements of businesses with mixed levels of commitment. But the German experience takes the next step: green politicians are now part of a ruling regime, they are sharing government in one of the most economically powerful nations on earth

Becoming the governors of a society has created new opportunities and new hurdles. The more radical messages of the Greens, formed in dramatic and direct opposition to ideas of unrestrained use which still fuel much of advanced industrialism, now have to be partly recast into the mould of workable government policies. Although environmentalists are still regarded as outsiders in many parts of the world, the German experience has proved that green parties are not just involved in single-issue politics. Rather, green politicians are capable of providing policy platforms which cross the full range of government portfolios: from economics to ecology; from development to conservation; and from the maintenance of peace to military activity.

Sharing in government has sorely tested certain fundaments of green ideology. The old tensions between the 'fundis' and the 'realos' were well and truly evident during Nato's bombing of Kosovo. The pacifist traditions run deep in European green movements, and the idea of Joschka Fischer, leader of the German Greens and the new Foreign Minister, endorsing the use of the Luftwaffe for the first time since World War Two to bomb Yugoslavia proved too much for certain supporters of the party. These tensions flared in many European cities during March/April 1999. 'How could one of the "four pillars" of Die Grünen – non-violence (alongside ecology, social equity and participatory democracy) be so easily jettisoned?' 'Was this really the type of trade-off necessary to govern?' These and other such questions dominated speeches at one such rally in Munich that the authors attended during the bombing. Indeed, it is this tension which is symbolised in Plate 1 and Plate 6 of this edition.

Alongside such trade-offs, government has also brought power to the German Greens to formulate and implement environmental initiatives like never before. Although not without its environmental critics, a raft of creative green legislation and administrative regulations is now emerging in Germany, and amongst some of its neighbouring European partners. Realistic timelines have been set in regard to the phase-out of nuclear energy in Germany; strict guidelines have been set in place in relation to genetically modified foodstuffs (GMO); incentive schemes have emerged to promote alternative energy systems; limits and audits have been created in relation to fossil fuel usage, with lower overall emission standards projected; and the use of chemicals in agriculture has been severely curtailed and monitored. These policies are just some of the more obvious outcomes of green government.

The German experience is exceptional. What makes the German policies all the more remarkable is that they occur within the well-recognised pressures of globalisation, which are often seen as promoting free market economies and ecologies, often regarding nation-state regulations and legislation as unnecessary 'interference' in resource production and allocation. In most parts of the world, environmental concern still has not been adequately addressed by either governments or business, and there is strong evidence that direct conflicts are actually increasing on these combined issues of globalisation and environmental degradation in both the majority and minority worlds.

This second edition shares the same goal with the first. In this book we have sought to explore the many ways of doing environmental politics and to consider the different ways in which power is used to either damage or protect the environment. The contrasts are great, but it is by considering the contrasts and similarities across the globe that it is possible to come to some sense of what is going on and what is at stake in the political struggle for our environmental future.

Tim Doyle
Doug McEachern

Preface to first edition

Environmental politics has many different faces. It is a world of contrasts and comparisons. Consider, for example, the contrast in a single day in Cape Town, South Africa. In the morning, early, we rose and drove out towards the east coast to visit Rondevlei Nature Reserve. This was set up in 1952 to conserve a wetland and its birdlife. It now lies behind a large mesh fence separating it from the 'coloured' area of Grassy Park and a black settlement. It is a good example of old-style wildlife conservation and, in its way, symbolic of the apartheid era. The focus of Rondevlei is well stated in its publicity brochure.

Although Rondevlei's original purpose was to conserve the rich birdlife associated with this important wetland, the emphasis has shifted to the conservation of the indigenous flora, fauna and natural features of the area, while still offering people a place where they can enjoy nature.

Rondevlei has also become an important environmental education centre, being within easy reach of the many schools in and around the Peninsula. It is a good bird sanctuary, preserving the habitats against urban encroachment and rescuing habitat from invasive exotic plants. There is even a population of hippopotamuses, although these are not often seen by the casual visitor. The emphasis is on wildlife and nature conservation and the people of the area come as visitors, outsiders 'to enjoy nature'.

In the afternoon, we travelled out to Langa, one of the oldest black townships, and visited the Tsoga Environment Centre. Langa was shaped and blighted by the power of the state and the apartheid imagination but possesses a civic spirit built on the pride of the African population in resistance to that malignant vision. Here in some recycled shipping containers is a thriving civic organisation dedicated to the greening of Langa. This group has developed a permaculture garden, and sponsored the development of local vegetable gardens and the planting of flowers on the verges and front gardens in the centre of the township. Tsoga confronts environmental degradation where it hits the everyday lives of the citizens of the new South Africa. Not everyone in Langa is enthusiastic about the goals of the group, as good housing and employment are the most pressing needs of the local community; but with great skill, commitment and dedication Tsoga is slowly making a difference, planting the appropriate trees to lessen need for the sangomas to collect healing plant materials from the wild and creating parks where before there were dusty open spaces.

Rondevlei and Tsoga show two different ways of responding to threats to the environment and human well-being. In this book we have sought to explore these and many other ways of doing environmental politics and to consider the different ways in which power is used to either damage or protect the environment. The contrasts are great, but it is by considering the contrasts and similarities across the globe that it is possible to come to some sense of what is going on and what is at stake in the political struggle for our environmental future.

Tim Doyle
Doug McEachern
April 1997

Acknowledgements

Tim and Doug would like to acknowledge Sophie Green for her excellent work as our research assistant for this third edition. We would both like to thank the students of the various courses we have taught on environmental politics at numerous universities, in different parts of the globe, for their demand for better-quality information and argument on these topics, and our colleagues who encouraged us in this project. We would like to thank Sarah Lloyd, our editor for the first edition, for her assistance, which went beyond that normally expected of an editor. Likewise, for subseqent editions, we must thank Andrew Mould and Jennifer Page of Routledge. We would also like to express our appreciation for the efforts of all those involved in environmental struggles, continually pushing forward their understanding of how environmental damage is caused and cured. It was their arguments and efforts which made a text such as this both necessary and possible.

Introduction

Although environmental issues and forms of environmental concern have a very long history, awareness of the environmental consequences of economic development was given an increasingly political character from the 1950s onwards (Young 1990). Individuals produced provocative studies warning of particular threats to the environment, as with Rachel Carson's well-known criticism of the increased use of DDT as a pesticide. Groups formed to press for solutions to particular or local problems or sought to get the political system to respond. Think-tanks, such as the Club of Rome, published accounts dramatising the potential depletion of the Earth's resources. International agencies, including the United Nations Environment Programme, began holding international conferences and promoting detailed studies of issues as part of an effort to get more co-ordinated and effective responses to increasingly global environmental problems. Later, protest movements, linking up with late 1960s student radicalism and with various anti-war mobilisations, took to the streets and forests in efforts to get a political response. In some places the mainstream political parties began to respond; in others, environmental concern was mocked and marginalised. In West Germany, a history of radical protest in the midst of obvious environmental problems and the nuclear threat of an active phase in the Cold War produced the formation of a radical Green Party (Die Grünen), which from 1980 to 1982 had enough electoral support to be represented in various state parliaments and in 1983 to enter the Federal Parliament in Bonn (Young 1990: 171). The image of the German Greens strolling into parliament in casual clothes, carrying potted plants and announcing that they were there to represent the politics of life was a sufficiently dramatic symbol to suggest that new forms of environmental politics were starting to challenge routine assumptions about the workings of the 'normal' political process. The post-unification Greens governed Germany in partnership with the SDP until only recently.

The purpose of this book is to provide an outline of the concepts that can be used to analyse and assess the character and consequences of environmental politics, in all its varied forms. In the course of presenting these concepts and the arguments about them, a broad outline of the development of these forms of environmental politics has also been provided. Most books on the environment begin with a litany of environmental problems and issues, but the very concept of the environment has been used in so many and in such contested ways that its meaning has become quite problematic. What counts as an environmental problem depends on varying ways of judging the ecological consequences of any particular act or development. It is important to understand that the term 'environment' is often constructed differently in different cultures and is used in different ways by different people. Not only is the word defined differently, but alternative clusters of issues are identified and a whole range of different kinds of politics can be generated on this basis. For example, environmental politics has been used to challenge the *status quo* in many societies. In North America, Australia and parts of Scandinavia the environmental agenda has often been dominated by attempts to protect wilderness areas from the intrusion and excesses of human development. Here environmental conflict has challenged the dominant goals of advanced capitalism and industrialism, such as unlimited growth and the rights of private property. It would be wrong to view environmental politics simply as a challenge to capitalist orthodoxy. In Eastern Europe, environmental politics developed as a rejection of state socialism's promotion of rapid and ecologically damaging forms of industrial and agricultural development. As part of the East European 'velvet' revolutions of the late 1980s, environmentalists championed pluralist democracy and 'free-market' economic solutions derived from the tenets of capitalism, in a bid to overthrow decades of rigid, bureaucratic and authoritarian rule.

Some Asian, African and South American countries have used environmental debates to challenge a different and global *status quo*, where a few affluent and powerful countries have access to, and consume, disproportionately greater amounts of the Earth's limited resources. The governments of these countries are less concerned with the rights of 'other nature' and are more concerned with promoting economic development to raise the standard of living of their populations. In these countries, environmental politics often focuses on issues of human survival such as the adequate provision of shelter, water, food sovereignty, equitable trade, environmental justice as well as safe work and healthy living conditions.

As far as these governments are concerned, the more powerful nations are busy promoting their own definitions of environmental ills, and promoting their own plans to solve these problems. 'Global ecology' is an excellent example of the environmental agenda of the 'developed world', since the global ecological issues on top of the North's environmental agenda are often not the same as those issues espoused by governments and peoples of the South.[1] For example, population control, species extinction, global climate change and deforestation are high-priority problems as defined in Northern elite and scientific terms. In the South, on the other hand, the traditional emphasis has been on solving environmental problems that have an impact on basic levels of standard of living and quality of life.

Governments of some industrialising countries oppose moves from the USA and Europe to impose global environmental objectives on them. Hence, in the international debates over greenhouse problems, Malaysia has set itself firmly against any moves that would make it harder for it to industrialise and export its goods. It should be noted that within these industrialising/developing countries environmental issues are still raised by groups of environmental activists. For example, in Nigeria, as we have already touched upon in the preface, the struggle of the Ogoni people against the military regime had a strong environmental theme. Here, poor people protested against the damage being done to their land by the polluting activities of Shell Nigeria and the links between Shell Nigeria and the oppressive military regime. In Indonesia, environmental groups oppose the logging of rain forests; in India, environmentalists oppose excessive logging of forests, the polluting consequences of industrialisation, and the environmental damage associated with population growth and urbanisation, and they are concerned about the consequence of global climate change. Further, in many parts of the world environmental politics has been used to contest inequalities and differences in power based on gender, class, race and species.

In the past ten years, since the writing of the first edition, there can be no doubt that majority and minority world environmentalists – whether acting within governments or outside them – still exist within distinct and different social movement traditions, but there is an increased interplay between different regions across the globe, creating a *green public space* which has become more transnational in character. As the preface to this edition explains, there is an increasing interplay between Southern and Northern green agendas. For example the South, while for sometime rejecting climate change as an affluent world side-show, is now engaging in climate debates, but reconfiguring them to focus on subsidiary issues such as *climate* debt (owed through the processes of ongoing colonisation) and associated nomenclature such as *climate justice* and *climate refugees*. The global South has attempted to incorporate Northern concepts such as climate, in part, in their efforts to explain their plight in affluent-world marketplaces, and there are some undoubted advantages to pursuing this path. But reconstructing all the diverse categories of the politics of environmental concern to fit under one climate banner also has it downside. The *climatisation* of southern green agendas may also confuse relatively simple issues of distributive justice: taking the direct conflict out of battles between *haves* and *have nots* and replacing it with more polite, less *political* frameworks and narratives.

This leads us to another key point. Although the concept of 'the environment' is invoked to support the struggles of the less powerful it can also be used by those either in or with power. It must be understood that environmental symbols are also used by the powerful to push their own interests. For example, some powerful business interests have also managed to enter the environmental 'corner', sometimes arguing that a healthy environment is one and the same as a healthy, expanding economy, with a marketplace unfettered by governmental controls and national boundaries. In fact, with 'sustainable development' a 'good' environment is actually 'good for business'. In this way, much environmental politics becomes oriented around efficiency and effectiveness criteria, consolidating the power of 'business as usual' or even, on occasions, 'business better than usual'.

So environmental politics is not just about 'goodies' versus 'baddies'. This symbol *environment* has such power that numerous cultures, and the powerful and powerless within them, invoke its name for disparate purposes. The definition of environment used may determine whether or not one is willing to accept the existence of an *environmental crisis*. For example, some environmentalists would argue that to speak of a 'crisis' is an unproductive, meaningless exaggeration. Environmental problems are not systematically linked and it is appropriate to make incremental adjustments to normal operating procedures as a response to evidence of particular pieces of ecological damage. In this way, 'sustainable development' constructs all environment 'problems' as efficiency issues, which have to be *managed* more effectively. These management issues include being more technologically, economically, organisationally, educationally and politically efficient. By constructing 'end-of-pipe' technologies to prevent waste products entering the environment as pollution or by redesigning the production process, these efficiency problems can be largely resolved. On the east coast of the United States, the 'environment' is almost entirely constructed in terms of 'air', 'land' and 'water' issues, and understood in these terms. Incrementally, through adequate recycling technology, among other things, all environmental problems can, in time and with effort, be solved. Greg Easterbrook (1995) is one who argues that all the hard decisions have already been made and from now on we just need to be better managers. He argues that:

- In the Western world pollution will end in our lifetimes.
- First World industrial countries are cleaner than developing countries.
- Most feared environmental catastrophes, like global warming, will be avoided.
- There is no conflict between the artificial and the natural.

Another very well-known and more recent author who builds upon Easterbrook's 'positive' view of the current state of the planet is Bjorn Lomberg in his book *The Skeptical Environmentalist* (2001). Like Easterbrook, Lomborg argues that trends are now looking increasingly favourable for the planet in the long term. He argues, for example, that the loss of forests has been more than adequately offset by improvements and increases in monoculture tree plantations grown specifically for logging; and that costs associated with risks like climate change are too high to justify action (Howes 2005: 19). Lomborg is consistently critical of environmental NGOs which claim that 'things are getting worse'. In fact, both Easterbrook and Lomberg dismiss those critics who suggest that there are serious environmental issues that cannot be addressed through 'environmental best practice' management as 'Cassandras'.

Often these environmental critics do use the notion of environmental crisis and portray the Easterbrooks and Lomborgs of the world as 'Pollyannas'. The whole debate between Pollyannas (overly 'positive') and Cassandras (overly 'negative') hinges on what depiction/construction of 'the environment' is being projected. The sustainable development 'Pollyanna' sees 'the environment' as external to human beings, an instrumental resource to be used and managed for human purposes but with a capacity to have impacts upon people and vice versa. It does not see humanity as intrinsically part of the environment or nature. Where humans are seen as 'intrinsic' to nature (by the Cassandras), environmental problems appear somewhat larger and more crisis-like. Issues include poverty, homelessness, disease and diminished diversity in terms of languages, lifestyles, ideas and political forms, as well as the loss of diversity in 'other nature'; the increasing gap between the haves and the have nots; the poisoning of the Earth through chemicals and other toxins; the extreme degradation inflicted upon the Earth by all versions of advanced industrialism, and the lingering threat of nuclear disaster.

There is a multiplicity of green political responses to the myriad of environmental problems around the globe. With the need for environmental problem solving to be both global and region-specific, some patterns emerge pointing to international solutions. In producing a comparative account of environmental politics, there is a tendency to replicate the assumptions and limits of the nation-state. Examples are drawn from events within particular countries and the similarities and differences between them can be noted and assessed. In considering the nature of environmental politics, it is important to notice the extent to which the division of the world into competing or co-operating nation-states is a problem with implications for understanding and responding to environmental problems. Often, an environmental problem has either a global/international dimension or its effects and solutions cross national boundaries.

A few simple examples can illustrate this point. If forests are cleared in Nepal, this affects the volume of water that runs off the hillsides into the rivers. An additional burden of silt from the consequent soil erosion results in greater flooding in Bangladesh. This is an environmental problem that requires co-operation across a national boundary for its resolution. Without successful international diplomacy, the environmental problem cannot be solved. This cross-border dimension may be present in a whole range of situations. Sulphur emissions from coal-fired power stations in Britain may be linked to acid rain in Europe and increased environmental stress on native forests. Other environmental problems are global in a more significant sense. The release of ozone-depleting substances into the atmosphere from a whole range of places across the globe can cause a change in the upper atmosphere that is then experienced as an environmental problem in a number of countries. No single country can solve the problems by its own actions since its individual contribution will be insignificant if other countries continue the unrestricted use of ozone-depleting chemicals. The enhanced greenhouse effect generates a similar global environmental problem.

Global solutions cannot simply emerge from the administrative documents generated by international diplomatic environmental forums. Green political solutions are partly to be found in the experiences of different political cultures. Equally important is the knowledge and respect shown by international environmental policy makers for these national and local traditions.

Scope and structure of third edition

Working in the United States in the mid-1990s, we began writing the first edition of this book. The task of selecting material for the initial edition was a selective, subjective process; but, to some extent, at least, it did broadly reflect the dominant literature on environmental politics at that stage. Most works on green politics had then only focused on minority world societies: usually 'Western', liberal-democratic and relatively affluent. The reasons for this are many. First, it is easier – particularly for minority world scholars – to study the politics of their own societies, due to institutional, industrial (particularly the insurance industry), funding, cultural and ethical constraints. Also, green politics had then been labelled (and still often suffers this fate) almost exclusively as *post-materialist* (particularly in the contexts of Australia and North America) and, therefore, was not usually seen as central to the concerns of less affluent societies.

There have also been profound changes since the 1990s, which must determine, to an extent, the subject matter of this third edition as we respond to 'our political times'. Two trends are worth commenting on here: increasing globalisation; and a perceived counter-trend towards authoritarianism. With globalisation, an associated key story in terms of environmental politics is the rapidly expanding importance of the global South in determining new constellations of ideas and practices within disparate environmental paradigms which relate directly to the politics of colonialism and post-colonialism. Global environmental politics, therefore, is increasingly wrestling with issues of distributive justice which can only be understood by appreciating historical inequities in resource distribution amongst nation-states due to hundreds of years of exploitation and colonisation. The case of Nigeria highlighted on the cover of this book, and featured in the preface, remains particularly salient here.

From the 1970s until the first half of the 1990s, it was widely imagined that a global trend had emerged: a movement away 'from dictatorial rule towards more liberal and often more democratic governance' (Carothers 2002: 5). This was considered a 'natural trend' – that everything would be 'alright' as global political communities became more homogenised to reflect the dominant pluralist values and system of the United States. But, the fact of the matter is, authoritarian practices are now re-emerging not just in transitional and/or one-party states such as Hungary, Nigeria or China, but sometimes within what are nominally considered liberal-democratic states. One reason for the resurfacing of authoritarianism may be corporate-propelled globalisation. Due to market globalisation 'democratic leaders often resort to authoritarian practices in order to secure market friendly results' (Skene 2003: 196). Obviously, this is not to say that these societies have become authoritarian at systemic and institutional levels; rather, elected governments can behave in an authoritarian manner in their acquiescence to the interests of private capital. As a consequence, although the first two editions did review the concept of diverse political regimes, this third edition must throw a net more widely to investigate environmental politics under non-liberal democratic regimes. Particularly in the case of societies living under the rule of authoritarian regimes, it was often assumed previously that green concerns were even more peripheral to the cut and thrust of everyday politics and survival. This assumption, of course, takes an overly restrictive view of what green concerns may entail in such countries.

Let us now turn to a brief outline of the structure of the book and its order of exposition of this edition. First of all, the basic structure of the book remains the same with one exception. The last decade of globalisation has been so profound in its impacts upon the politics of the globe that the structure of the book needed to be changed. In the first two editions, we finished the manuscript with a final chapter on 'the global dimension of environmental politics'. But the forces of globalisation cannot now be separated out into a convenient chapter at the end: there are global dimensions of social movements, NGOs, corporations, administrative and institutional arrangements and even in green political parties. So now we have decided to incorporate these global implications into each chapter dedicated to a particular political form as it wrestles with the politics of environmental concern. Thus, the *global dimension of environmental* politics is now treated as an important feature in all chapters. Chapter 1 outlines what is meant by the central concepts of politics and environmental studies. The different possible definitions of each concept vastly affect relations between the two, shaping what needs to be understood and the terms in which the analysis will be conducted. Power is basic to both the practice and the assessment of politics. This chapter presents a series of different definitions of power, the models of society on which they are based and the different methods they suggest for the study of environmental politics, conflict and policy making.

An important section within this chapter has been substantially expanded. Most of the previous editions have been based on environmental politics in liberal democracies. Although in the first two editions we did address the differences between liberal democracies, military dictatorships and one-party states, this description will be expanded to include several case studies based on recent research in Iran, Burma and Thailand. Increasingly, it can be argued that green politics is emerging under authoritarian regimes (and even within certain liberal democracies and within large transnational corporations). The book now addresses this growth in green politics outside of mainstream, Western models of politics. Also, at the end of the chapter new key study terms are introduced such as: environmental justice, environmental sovereignty and security, and environmental citizenship.

Chapter 2 provides a summary of the key versions of these political responses and their worldviews. We provide three different frameworks for understanding a cacophony of political ideas which gather under the environment rubric. Importantly, the use of each frame dramatically alters the understanding of these ideas, and how they relate to each other. The first frame is one which we used in the first two editions. There have been three broad types of political response to environmental concern: to reject and resist it; to accommodate it and propose reform; or to embrace it and demand revolutionary or radical change. Each of these responses is associated with an ideology or 'world view' made up of ideas and value judgements. Sometimes these ideologies are coherent and cleanly structured; on other occasions they are a jumble of ambiguous and contradictory values and beliefs.[2] The second frame we introduce is the more traditional left–right political axis. This is done for two major reasons. First, many variants of modern environmental thought are really intersections with traditional political thought. We need to understand the basic principles of, and differences between, for example, liberalism and neo-liberalism, if we are to garner a rich understanding of concepts of a responsible governance versus market-based responses to environmental concern. Also, the left–right axis, far from drifting from view in the polities of the South, has continued currency in debates about a range of issues, from resource distribution to carrying capacity.

In previous editions there was a sustained focus on Western ideological traditions and their green connections. This made sense, as modern environmentalisms first emerged in Western societies. Still, there is a need to understand basic differences between philosophical positions within Western political thought, like the attributes of social ecology versus deep ecology, and the manner in which diverse forms of feminist thought intersect with green ideas. In this new edition, however, in a bid to further globalise the material, we also look at the way in which diverse religions and philosophies in both eastern and Western cultures interface with green ideas. For example, the ideology of sustainable development is interpreted very differently in China from its interpretation in previously industrialised, Western countries. China is currently undergoing a phase of unprecedented and profound industrialisation. In the Chinese case, predominantly Western green concepts such as sustainable development and *Agenda 21* are wedded to both Confucian and post-revolutionary thought. The outcome is a profoundly different green ideology from that experienced elsewhere. In effect, Chapter 2 is expanded to include ideological interests that drive environmentalism in the majority world (the less affluent world).

This book is conceived as an introduction to the comparative study of environmental politics. It is assumed that what happens in environmental politics varies between countries on the basis of the different political systems they possess, the different kinds of environmental problems experienced and the severity of those problems, the differing strength and capacity of governments to make and enforce environmental decisions, and differences in culture. The broad concepts needed to study environmental politics in any of these countries may be the same but their application will provide quite varied interpretations of why things have happened in the way that they have as well as contrasting evaluations of the significance of different outcomes. Both elements of similarity and differences will be stressed.

Apart from providing a comparative look at different nations' experiences with environmental politics, this book is also comparative in its investigation of different types of policy process within these countries. Policy is not a document but a political process. Numerous processes can be used to create, sustain or thwart environmental change. Often, political structures determine what types of environmental goals can be pursued. In addition, depending on what the environmental goals are, radical or reformist, certain political structures are more useful than others. For example, environmental politics is played quite differently in informal networks within social movements from within more formalised non-governmental organisations. It is played differently in political parties from in bureaucracies and administrative systems. It is played differently in business corporations from in elected governments. Each kind of organisation produces different approaches to environmental politics and each form interacts with other structures of politics and policy making. Chapters 3 to 7 presents accounts of the varying ways in which different political processes have responded to the 'greening of society'.

Chapter 3 presents different ways of analysing the general character of social movements, with particular attention paid to the specific attributes of environmental social movements. What are social movements? Why do they emerge in particular countries at particular times? How are they structured? What is their potential for promoting or resisting environmental change? What kinds of politics are associated with these new social movements? How does the form of politics vary in different parts of the world: from post-materialism, post-industrialism and post-colonialism? This chapter emphasises the importance of informal politics, which often cuts across national, cultural and institutional boundaries. Importantly, in this new edition, there is a final section on the character of transnational environmental movements. Connections are drawn and distinctions made between green movements and the rapidly emerging anti-globalisation movements: from the Zapatista movement in Central America, to European and North American anti-globalisation protest movements.

Non-governmental organisations (NGOs) are the subject of Chapter 4. NGOs are the most visible players in environmental politics around the globe. They are involved in many different spheres of politics, from the local community level through to the politics of the nation-state and international or transnational politics. They exist in both the predominantly non-institutional domain of social movement politics, and the institutionalised milieu of political parties, administrative systems, governments and beyond. There are many theories constructed around the correct understanding of NGOs, and usually, these theories depict NGOs differently due to the fact that each conflicting 'grand model' of politics on which the theorising is based, is rooted in very different understandings of the construction and role of the state in those societies. In short, your view of the state will alter your understanding about the role, function and operations of NGOs. In this new edition we look at four separate theories: pluralism, corporatism, authoritarianism and post-modernism. It has been necessary to add the section on authoritarianism due to the fact that NGOs are now appearing as legitimate political actors under one-party and movement states such as

Cuba, China, Burma and Iran; whereas once they were considered as operating exclusively in liberal-democratic domains.

Furthermore, we will explore the impacts of globalisation on NGOs, using the cases of northern-based, transnational NGOs, making a clear distinction between what we call emancipatory groups (EGs) versus those which are part of a green governance state (GGS).

Chapter 5 of the original and second editions looked at green electoral politics in four countries, all of which are Western democracies. This focus on green politics and the formation of green political parties in the liberal democracies of Western Europe, North America and Australasia was justified in the past as electoral responses which included environmental concerns had been almost exclusively a Western phenomenon, in multi-party states. This largely remains the case, and the bulk of this chapter continues to reflect this reality. Also, however, there have been attempts to incorporate three important recent trends. First, in recent times, a party-political response has emerged in non-Western democracies, such as evidenced by the fascinating new green party experiences in places such as, for example, Taiwan and South Korea. Second, for the first time, we have also seen one-party states – such as Cuba, China and Iran – having to confront and address the politics of the environment in serious fasion. Finally, even electoral politics is not immune from the processes of globalisation. In the past, political parties almost exclusively reflected the domestic political terrain within nation-states. The emergence of the Global Greens – a coalition of green parties around the world – challenges these preconceptions.

The bulk of the chapter is divided into three sections reflecting the three basic ways in which environmentalists have responded to electoral systems: by abstaining from electoral politics; by influencing existing political parties; and by creating new, green parties. By far and away to most significant factor in determining the shape of a country's electoral response to the environment relates to the type of electoral system it has in place. We review national examples of three systems: proportional representation in Germany, Sweden and Taiwan, the first-past-the-post system of the United Kingdom and the United States; and the preferential system of Australia. Within this framework consideration is also given to the various other factors that shape the development of green parties and allow them a degree of electoral success. The most significant examples are, of course, the case of the German Greens and the Green Party in

Sweden, but there are many other examples of green party formation and political success inside the standard processes of representative politics. This chapter also explores those factors that inhibit the development and success of these parties.

Many works that analyse politics pay little attention to the role played by organised business. It is sometimes assumed that politics takes place in a distinct public sphere and that business is confined to the economy; but on questions of the environment the business of the marketplace has a strong political dimension. Particular firms and various business associations can be active in defending themselves against various environmental campaigns and the threat of environmental regulation. At times, business has been the key player in trying to define an acceptable approach to the making of policy on environmental questions. Much of the content for the concept of sustainable development has come from the efforts of business and the research it has sponsored. This is even more true of the development and dissemination of 'free market' arguments about how to respond to environmental problems. Chapter 6 explores both the various power resources deployed by business and the strategies developed by different parts of business to protect themselves and the right to pursue 'business as usual'. Finally, this chapter will use a major case study as background to discuss the three types of business response to the environment and, in particular, to the issue of climate change. Some businesses have continued to reject the existence of climate change; some have accommodated it within existing practices in different ways; and some have developed a genuine environmental response, pursuing, for example, diversified and alternative forms of energy to those derived from fossil fuels.

Chapter 7 discusses institutional politics and policy making. When an environmental issue is put on the political agenda and government chooses to respond, that response will be enacted, regulated and monitored by the administrative arm of the state. It is not just that the bureaucracy does what the political process tells it, since important information and policy advice is generated by the bureaucracy itself. The very character of the policy response may well be determined by what the bureaucracy does. In recent years, there has been a major debate about the appropriateness and efficiency of direct governmental regulation of environmental problems as well as strong advocacy of the use of prices and markets to produce better environmental outcomes. These important questions are also considered in Chapter 7. These institutional responses range from every level of government, local, state and federal. At the local level we provide information as to how some local bureaucracies have responded to *Agenda 21*, an initiative which emerged from the first Rio Earth Summit in 1992. Whilst on the subject of international 'summits', its important to understand that institutional and administrative responses to the environment also cross the borders of nation-states, with many attempts over the past decades of creating international and transnational agreements between nation-states. More than that, some environmental issues cross national boundaries in the sense that action in one country produces environmental harm in another. Others are more global in character and require co-operative, negotiated, legalistic, institutional and diplomatic efforts to generate even plausible responses. As a consequence, this chapter explores what is needed to understand the success or failure of attempts to develop effective international administrative regimes to deal with these complex global and cross-border environmental issues. It also considers a number of case studies to show what has been done in various major international forums, such as the Rio Earth Summit of 1992 (and the subsequent 'Rio + 10' conference), and in the attempts to develop treaties in response to global issues, such as population control and hazardous waste disposal. Leading on from the last chapter, the role of large transnational players such as the World Bank and the WTO are reviewed here.

The book concludes with a survey of the broad analytical concepts required for a comparative understanding of environmental politics as well as a restatement of the importance of the comparative perspective. Furthermore, the emergence of the dual concepts of environmental security and environmental refugees will be articulated. In the ten years since writing the first edition, the politics of environmental concern, if anything, has assumed more gravity and importance.

Further reading

Connelly, J. and G. Smith (1999) *Politics and the Environment: From Theory to Practice*, Routledge, London.

Doyle, T. J. (2005) *Environmental Movements in Majority and Minority Worlds: A Global Perspective*, Rutgers University Press, NJ, New York and London.

Dryzek, J.S. (2005) *The Politics of the Earth: Environmental Discourses*, Oxford University Press, Oxford.

Easterbrook, G. (1995) *A Moment on the Earth: The Coming Age of Environmental Optimism*, Viking Penguin, New York.

Howes, M. (2005) *Politics and the Environment: Risk and the Role of Government and Industry*, Allen and Unwin, Sydney.

Lomborg, B. (2001) *The Skeptical Environmentalist: Measuring the Real State of the World*, Cambridge University Press, Cambridge.

1 Politics and environmental studies

- Central concepts of politics
- Political regimes
- Dimensions of power
- Models of policy processes
- Key study terms

Introduction

Both environmental studies and the discipline of politics can be conceived of, and practised, in quite different ways. For example, environmental studies can be treated as a sub-discipline of natural science and bring to bear that whole approach to understanding particular problems and to evaluating alternative solutions. That is not the only approach. It is equally possible to understand environmental studies as essentially an interdisciplinary enterprise bringing together knowledge and expertise from a whole range of scientific and social science disciplines. The study of politics is capable of equally varied conceptions, ranging from its basic focus through to questions of method and means of evaluation. This chapter starts with a consideration of the delineation of the two fields of environmental studies and politics to establish what is at the core of the comparative study of environmental politics. On that basis some of the key, even if contested, concepts of politics are outlined along with sketches of alternative methodologies and assumptions about the workings of the different political processes found across the world today.

Defining politics and environmental studies

What is environmental studies?

Environmental studies can be approached in a number of significantly different ways. It is possible to treat the field as being coterminous with that of 'ecology', the study of the particular ways in which living matter combines in changing and stable patterns (Wilson 1992; Flannery 1994). Here ecology is a branch of science to be approached with the same methodological assumptions as other areas of scientific inquiry. Such an approach could generate studies of, for example, the ecology of a particular region (the redwood forests of North America, the moorlands of England, the Great Barrier Reef in Australia, or the Karoo in South Africa). Even in such studies of nature and natural systems there would be some recognition of the part played by humans in this ecology. For example, human actions may destabilise a particular ecological balance and impose new, managed stability/instability, and this could be noted with the same kind of scientific detachment as for the study of other disturbances.

Recognising the human element indicates one of the problems for defining environmental studies only in ecological terms. How are the actions of humans and human societies to be interpreted? It is certainly possible to invent a 'human ecology' and treat this as a sub-set of other parts of scientific study: it could form part of zoology, biology or human geography. It is certainly possible to gain some understanding of the character of environmental problems by considering population dynamics and the causes and consequences of 'overpopulation', but how good or useful is this interpretation? There is more to understanding the impact and dynamics of human social organisation than can be generated from this particular perspective. Here it is possible to extend the conception of environmental studies and to enhance the interdisciplinary element. Ecology itself involves forms of interdisciplinary study. To understand the interaction of life forms means crossing, at the very least, the boundaries between botany, zoology and soil science. But this interdisciplinary focus needs to be enhanced to include, at the least, the social sciences if we want to understand and assess the environmental consequences of humans as socially organised.

From this perspective, environmental studies needs to combine understandings from both the sciences and social sciences and to be interdisciplinary in this sense. Knowledge derived from ecology and a whole range of individual disciplines should be combined with knowledge derived from the social sciences. For example, human impacts on their environments depend on a whole range of factors: cultural attitudes to nature; modes of social and economic organisation; and the kinds of political process that can either protect or harm the environment.

Plate 1 ***Protest march in Munich against bombing of Yugoslavia by Nato forces, March 1999. Author's private collection***

Depending on how environmental studies is defined, the form of political knowledge required will be quite varied.

What is politics?

What is understood as 'politics' and 'political' varies widely. Often politics has been defined in a particularly narrow way and the word is used to refer to processes of government; decision making and administration; elections; the machinations of political parties; and the efforts of groups to influence these political processes. This limited, 'government-centred' view of politics, according to Crick, emerged in advanced, complex, usually European, societies:

> The establishment of political order is not just any order at all; it marks the birth, or the recognition, of freedom. For politics represents at least some tolerance of different truths, some recognition that government is possible, indeed best conducted, amid the open canvassing of rival interests. Politics are the public actions of free men.
>
> (1964: 18)

This view of politics, which sees government as a public instrument of freedom, is associated with a 'Western' tradition, reaching back to the ancient Greeks and illuminated in the writings of Aristotle and Plato. Here politics is seen in a positive light, as part of the way in which citizens are fulfilled and the highest goals of a community are achieved. Involvement in public debate over common problems and their solutions was seen as part of civic duty: to be involved in politics was a high cultural and social ambition. In essence this is a conception of politics as some variant of democratic politics. Today, democracy and democratic politics are greatly valued but the practice of politics has separated out the role of citizen from that of the politician. Politicians are professionals elected to act on behalf of voters in remote parliaments and governmental processes. It is largely their job to reconcile conflicting and convergent societal interests through compromise and negotiation in a public sphere. In this view, politics exists 'out there'. Most citizens are removed from its daily reach. Politics is regarded as that peculiar set of relationships forged in the parliaments and congresses of national and state capitals. It is important to note that Crick's limited view of politics is shared by most people. Given this alienation, politics is often seen in a negative light, as something tawdry, deceitful, largely a battle between arrogant and egotistical men and, worst of all, a waste of time. Nonetheless, a 'narrow' view of politics provides workable boundaries for political enquiry.

Critics of this perspective see politics in broader terms, as far more universal, capable of crossing cultural boundaries and existing within and outside the institutional boundaries of the modern state (and outside Europe!). Politics is not just confined to the actions of government but is also found in the so-called private sector of the business 'community' and in the more informal realms that often operate outside the state. In fact, Leftwich argues that politics exists 'at every level and in every sphere' of human societies, and that political

> activities are not isolated from other features of social life. They everywhere influence and are influenced by, the distribution of power and decision making, the systems of social organisation, culture

> and ideology in society, as well as its relations with the natural environment and other societies. Politics is therefore the defining characteristic of all human groups and always has been.
>
> (1983: 11)

Politics, in Leftwich's terms, occurs in homes, sporting clubs, workplaces and in the street, as well as in parliaments. A strength of Leftwich's inclusive depiction is its capacity to analyse non-institutional politics. In this way, politics refers to our relationships to one another and our interactions, in many different collective and sub-cultural forms: as individuals; as members of families; as informal networks and groups; as organisations; as governments; as corporations; and our activities in a whole range of other institutionalised settings.

All definitions of politics are contested and value-laden. There is no one universal definition. It is important for students of politics to be explicit about the reasons used to justify one definition over another. Either the narrow or the broad definition of politics could be used to frame a consideration of environmental politics, but quite different studies would be produced. Use of the narrow definition would give an account based on the constitutional design of the political process, its institutional characteristics, and the role of political parties and pressure groups as well as a consideration of the way in which environmental policy is made and administered. And, indeed, these topics are covered in this book.

A broader definition would lead to the inclusion of non-institutionalised forms of environmental concern and activism, which are never simply captured by an organised mobilisation into normal politics. Attention is also directed towards the political debates and conflicts that occur within these informal movements and in the more non-institutionalised settings. These, too, are included in this study. A narrow definition keeps more things out of consideration, and hence makes the task of analysis more manageable. A broader definition brings in more things that need to be included but at the risk of a more diffuse and unmanageable study. There are costs involved in the use of either definition: it is good to be aware of what is gained or lost by choices made about preferred definitions.

What is the place of politics in environmental studies?

> Environmental studies' . . . essential interdisciplinarity is intellectually testing and touches not only on biology and ecology, but all the social sciences, most especially politics and economics.
>
> (Doyle and Walker 1996: 1)

It is tempting to say that politics is just one discipline that should be incorporated into environmental studies, but far more is required than this: politics is central to environmental studies. The relationships between the two differ, depending on the definitions of each. Like politics, environmental studies is defined in numerous ways to include and exclude various modes of inquiry. Many of these differences emerge around conflicting views as to exactly what the environment entails. For example, if the environment is seen as a biophysical reality existing outside of humans, then environmental studies is the study of the relationships between human and non-human worlds. Often this 'external' view of the environment gives birth to instrumentalism; that is, the environment (or nature) is understood as a series of resources (natural products) or a series of processes, which need to be managed by humans in a renewable, sustainable, manner. Doyle and Walker argue as follows:

> environmental studies is fundamental in the fullest sense. Human existence depends completely on the continuing availability of inputs such as air, water, foodstuffs and other resources from the natural environment. Without ecology there can be no economy and no society.
>
> (ibid. 1)

This instrumental depiction of environmental studies fits snugly with many definitions of politics. For example, Leftwich's writings on politics portray the interface between humans and 'natural resources' as a centrally defining part of politics. He contends that

> the major organising activity at the heart of this history of cooperation, conflict, innovation, and adaptation in the use, production and distribution of resources has been and still is *politics*.
>
> (1983: 12)

As society is seen as outside the environment in this view, the study of politics complements the natural sciences (and the many other disciplines involved in environmental studies) in allowing us to understand and manage our biophysical resources; to manage our environment; to formulate and to implement environmental policies.

The role of politics in environmental studies increases in scope quite dramatically if we consider a different notion of the environment where humanity is part of nature and not distinct from it. Nature is no longer seen as a resource but as a more holistic construct that is far more open to interpretative possibilities. Being considered part of nature, there is a recognition that our relationships with the non-human world are socially, as well as (on occasions) biophysically constructed. So environmental studies now includes those relationships already mentioned between humans and non-humans, between differing non-human entities, and the relationships between humans themselves. Under this rubric, politics meets environmental studies in different ways. For example, issues of social democracy (participatory and representative), non-violence, social equity and justice as well as ecology now dominate the intellectual and green activist agenda. New types of relationship and entire human and non-human societies are analysed and imagined based on this more integrative, non-anthropocentric view of the environment.

Concepts for the analysis of environmental politics

Political regime and environmental politics

The particular way in which the political system is organised has a strong impact on the scope and effectiveness of environmental politics. There is a whole variety of different political regimes, and different ways in which rulers and ruled are connected and the institutions of politics are designed. The most significant impact political regimes usually have on environmental politics is at the level at which democratic participation in the political sphere is permitted within that regime.

As the nature and expression of social and environmental activism within a particular society is often determined or framed by the type of political regime under which it operates, it is important to understand the importance of the existence, or otherwise, of levels of democratic accountability and political openness in the societies we study. Broadly speaking we could use a progressively authoritarian schema where regimes qualify as liberal democratic representative, hybrid or traditional authoritarian (Doyle and Simpson 2006). Ideal liberal democratic representative regimes epitomise the democratic freedoms of association and thought with regular free and fair elections. Traditionally, authoritarian regimes have been military dictatorships or one-party states with a lack of both democratic freedoms and competitive elections.

In between these two extremes there are what we could call hybrid regimes which have burgeoned in number since the end of the Cold War. These regimes cannot easily be included in either of the main two categories as they maintain some democratic structures, such as elections, but they also restrict genuine political activity in this and other areas enough that they could not be considered liberal democratic.

There are many variations within these three basic categories and only some important elements will be discussed here.

Examples of the liberal-democratic representative political regime are to be found in North America, Europe and Australasia, with variations in Japan, India and South Africa. Here the link between rulers and ruled is based on elections and the vote. Each citizen has a right to vote and a right to take part in politics. Periodic elections are held where political parties compete for the votes of citizens. On this basis, government is both formed and legitimated. Such a government is authorised to act on behalf of its citizens until such time as another election is required.

This can be characterised by the label 'liberal-democratic representative regime' because of the three main elements combined within the system of rule. First, there are liberal elements, characterised by a bundle of rights and freedoms that are necessary for participation in this type of politics. These rights typically include freedom of speech and opinion, and freedom of assembly. Such freedoms can be embodied in a Bill of Rights (as in the USA), in a written constitution (as in France, Germany and Italy), or in the conventions and practices known as the common law in states without written constitutions (as in the United Kingdom). Second, there is the democratic element symbolised by the right to vote in periodic elections; that is, the right to have a say or determine which political party or parties will form the government. It is in this limited sense that the people can be said to rule or govern themselves in this type of regime. Third, there is the representative element. These are not political regimes based on direct democracy where the people have a significant or effective say in all aspects of everyday governance. Instead, the people rule in an indirect manner, through elected representatives who take part in government, if their political party is in office and if they are selected to serve as ministers.

There are many different variations within political regimes of this type. Electoral systems can differ greatly from 'first-past-the-post' in single-member electorates, where the person who gets the most votes wins (as in the United Kingdom), to forms of proportional representation with multi-member electorates (as in Germany), and to preferential voting where the winner requires an absolute majority constructed by progressively eliminating the lowest-placed candidates and distributing preferential votes (as in Australia). Such states can take a unitary form (as in the United Kingdom) or some form of federation (as in Germany, the United States of America, India or Australia), where legislative powers and responsibilities are divided between a central or federal government and the states or provinces. Again, there are many different ways in which federal systems can be designed and operated. Party systems, the form of pressure group activity, and the role of trade unions, business groups and religious organisations can all vary widely. A comparative understanding of what happens in environmental politics needs to take into account and evaluate the significance of these variations.

The openness of liberal-democratic representative regimes contrasts sharply with the closed nature of politics in authoritarian states such as those ruled by various forms of military dictatorships. Military governments come and go. When the first edition of this book was published, Nigeria, Burma and Indonesia could be cited as examples of the different kinds of this type of regime. Of the three only Burma remains subject to authoritarian, repressive military rule. Nonetheless, the military continues to rule in a number of states in Africa and, effectively, Pakistan. Characteristically, the military comes to power through a *coup d'état*, displacing a previous regime, which can have been either authoritarian or some form of democratic regime. Once in power these governments are likely to be repressive and to limit the role of opposition on environmental or any other grounds. Just because these are military regimes and rule in an authoritarian manner does not mean that politics comes to an end. The form of politics changes and the number of people who can be effectively involved is sharply reduced. The politics of opposition is also transformed so that much of its efforts to influence what is done is tied to arguments about the political form itself, often in attempts to introduce more open and more democratic political processes. For example, this can be seen in Burma with the struggle of Aung San Suu Kyi and her supporters against a harshly repressive military regime (see Box 1.1). Within the military there will be the standard jockeying for power and, at times, efforts at internal reform.

Box 1.1

Examples of three types of regimes

Military Dictatorship: Burma (since 1962)

Senior General Than Shwe has been the effective ruler of Burma since 1988.

1 Elections are not held, the media is tightly controlled by the state and the government is unaccountable.
2 Human rights and environmental protection are intimately linked.
3 NGOs and activists generally face harsh restrictions on freedoms and activities and face imprisonment and beatings for speaking out.
4 Independent NGOs may be tolerated if of the 'community development' type but proscribed if they challenge the political status quo.
5 NGOs and activists often appeal to international NGOs/agencies as they expect little positive response from the military regime (Doyle and Simpson 2006).

Hybrid regime I: competitive authoritarian: Thailand

Thailand under Prime Minister Thaksin Shinawatra

Sites of authoritarianism or contest

1 *Electoral*: Elections regularly held but may be characterized by large-scale abuses of state power, biased media coverage, harassment of opposition candidates and activists, and an overall lack of transparency.
2 *Legislative*: Weak relative to executive power. Unable to exercise effective parliamentary oversight.
3 *Judicial*: Governments routinely attempt to subordinate the judiciary, either blatantly or through more subtle techniques such as bribery and extortion.
4 *Media*: Executives try to suppress the independent media, using relatively subtle mechanisms of repression such as bribery, the selective allocation of state advertising and restrictive press laws that facilitate the prosecution of independent and opposition journalists. May also use 'proxies' to harass or physically intimidate journalists.

5 *Law enforcement agencies*: Police may be corrupt and in the pocket of business interests or politicians (or both), and harass environmentalists or social activists who oppose development policies. They may be immune from prosecution for human rights abuses and may engage in kidnappings and extrajudicial killings (see also Box 1.2).

Hybrid regime II: Iran (although since 2004 heading towards traditional authoritarianism): state theocracy Iran – Islamic revolution in 1979

1. Elections are held but there is also arbitrary rule by the Islamic Supreme Leader and Guardian Council who can veto legislation and who vet election candidates to ensure their candidacy does not undermine 'Islamic law'.
2. From 1997 to 2004 the reformist movement gained momentum in parliament under a reformist president.
3. In the 2004 parliamentary elections the Guardian Council banned 2,500 candidates from standing.
4. In 2005 presidential elections 1,000 candidates were banned from standing and a hardline Islamic conservative was elected.
5. No freedom of the press, no dissent tolerated (Doyle and Simpson 2006) (see also Box 4.4).

There is another sense in which politics continues inside these military regimes and it is surprising how often forms of representational politics are mimicked in these regimes. For example, almost as soon as the Suharto regime came to power in Indonesia, it set up a political 'party', Golkar, and various trade unions, business organisations and civic bodies to try to integrate the population into the regime. Opposition political parties existed, even if their leadership was, at times, determined by the government, and elections did take place. A similar process occurred in Chile when the Pinochet regime came to power after destroying the elected Allende government. These 'representative' organisations and processes provide some scope for the practice of oppositional politics, including politics based on environmental concern.

Just as with liberal-democratic representational regimes, there is considerable variation between these military regimes. They differ in terms of the means by and the circumstances in which they come to power. They differ greatly in the amount of repression used to maintain themselves in power. There may be a great deal of difference between the development strategies produced by these governments, although frequently economic development will be linked to strategies for gaining wealth for particular individuals and families. They will also differ in the scope they give to the press and other opportunities for public debate. Finally, they will differ in the degree of stability and security of their domination and many, as discussed earlier, have been abandoned in favour of hybrid regimes.

There is another set of authoritarian regimes that can be characterised in terms of one-party domination. In the recent past, many of these were the classic one-party states like the USSR, the People's Republic of China and the Democratic Republic of Vietnam. Here a party, claiming to be inspired by the works of Marx and Lenin, dominated the social, economic and political life of these countries, while vigorously pursuing economic development, often with a strong emphasis on rapid industrialisation. Once these regimes ran centralised, planned economies with rigid bureaucratic controls. Many of them (China and Vietnam included) have now 'liberalised' the economy but most still rigidly control the political process. Some have given up entirely and have been succeeded by hybrid regimes with varying degrees of openness. One-party authoritarian regimes and military dictatorships often seem to show family resemblances as control and use of the military is an important part of the ways in which these regimes rule.

Hybrid regimes can be formed by the disintegration of one-party states or military dictatorships but they can also be formed by liberal democratic regimes that are hijacked by an individual or group and thereafter experience a deterioration of democratic freedoms towards authoritarianism. An important example here is the Southeast Asian country, Thailand, which has a dynamic capitalist economy and, since 1992, a nominally liberal democratic political system. After a new, more democratic, constitution came into effect in 1997, Thaksin Shinawatra, one of the country's richest individuals, rose to become prime minister and presided over an unravelling of the new democratic measures to protect domestic business interests, forcing a movement back towards authoritarianism. As there are democratic and authoritarian elements in the regime it could be classified as 'competitive authoritarian' using the criteria in Box 1.1 (also see Box 1.2). It remains firmly entrenched in the hybrid category, however, with much greater civil and political freedoms than in a traditional authoritarian regime such as Burma's

Box 1.2

Thailand: a new hybrid regime?

Thailand has been a predominantly liberal democratic constitutional monarchy – like countries such as the UK and Australia – since a process of 'Westernisation' occurred in 1932. There have been periodic military coups where the military grabbed power but the last of these was toppled by popular protests and intervention by the king in 1992.

Throughout the 1990s social and environmental movements agitated for a new, more democratic, constitution and in 1997 achieved this goal. The new constitution aimed at providing greater public participation in development decisions and the running of government, and more checks and balances within government.

In 2001 the ascent to power of Prime Minister Thaksin Shinawatra, a Thai billionaire, and his party Thai Rak Thai (TRT) – which was dominated by wealthy capitalists – resulted in a rolling back of these gains as he used his political and economic power to undermine democratic institutions. Environmental NGOs and activists faced increased pressures such as police and military repression when opposing large-scale developments that caused environmental destruction or social dislocation. Environmental activists were killed by assassins, who were rarely caught, and occasionally, the police were also implicated in the murders. The independent media and the judiciary were also subject to intimidation from the government and related business interests with dissenting media often facing economic and physical coercion. TRT had a large majority in parliament and Thaksin progressively ignored the legislature, rarely making an appearance, and increasingly ruling by decree.

In 2006 public protests over Thaksin's authoritarian rule finally pressured him to resign but he retained the leadership of TRT and activists expected him to attempt a political comeback. In September 2006, however, the military has staged yet another coup, seizing the mechanisms of government again. The Thailand case study demonstrates that there are new *hybrid* regimes emerging, such as the *competitive authoritarian* regimes, that combine a capitalist economy and elements of liberal democracy with authoritarian tendencies which limit the freedom of speech and action undertaken by environmental movements (Simpson 2006). These periods of hybrid democracy are sometimes interspersed with times of military control.

(Simpson 2006). A military coup in 2006 further restricted the democratic elements of the regime.

In other cases hybrid regimes may come about through revolutionary change such as in the hybrid, but increasingly authoritarian, Iran. Since the Islamic Revolution in 1979, the Iranian government has combined some elements of democratic participation with arbitrary Islamic rule from above. Since 2004, the democratic aspects of the regime have suffered at the hands of the ruling elite and a conservative Islamic authoritarianism now governs the country (see Box 1.1). In all these cases there may be democratic and representational organisations and elections but they may have more form than substance. Other 'civil' organisations may lack independence and autonomy. Politics still happens in both hybrid and authoritarian regimes, shaped by the ideological claims of the regime and factional struggles.

The character of the political regime has an impact on the scope and effectiveness of environmental politics. In liberal-democratic representative regimes there is significant scope for those concerned about environmental issues to have their say and to try to influence the political process. People are free to form groups, to join political parties, to invent new parties and to go into politics to achieve their goals. None of this says that environmental activists will be successful. Indeed, significant social and economic forces, equally able to organise, will oppose them and seek to keep environmental issues off the agenda. Nonetheless, the design of the regime provides scope for environmental politics.

In authoritarian regimes, especially those committed to rapid economic development, there is very little scope for environmental politics, even by loyal supporters of the regime. However, environmental politics and actions can still happen. The struggle of MOSOP (Movement for the Survival of Ogoni People) over the fate of the Ogoni lands under the military dictatorship in Nigeria included a strong environmental critique of the operation of the Shell Oil Company as part of its claim for autonomy (Osaghae 1998: 246, 304–6; Osaghae 1995; Naanen 1995, Oswka 2005). Even in Vietnam and China, some environmental issues are recognised in official circles. Under hybrid regimes there is a greater, and sometimes bewildering, mixture of democratic freedoms with authoritarian restrictions. In the case of Thailand, there is a formal and institutionalised Environmental Impact Assessment (EIA) system with compulsory public hearings, but the regime has, in some cases, prevented opponents having access to the hearings through the use of police roadblocks and wire barricades.

Students of environmental studies need to be aware of these different kinds of political regime since they have an impact on the fate of environmental politics. They also need to be aware that there are other kinds of regime, other ways of linking rulers and ruled, and that these too have different consequences for the ways in which environmental politics can work.

Dimensions of power

Power is involved in all environmental conflicts and policy making. Both doing environmental harm and protecting, conserving or saving the environment require an effective deployment of power to prevail against opposition and power to make and enforce decisions. Unfortunately, the very terms needed to understand the character and consequences of power are contested. Both academics and political activists are divided in their views on what constitutes power, how power should be analysed or can be effectively deployed and how to assess the legitimacy of the ends of power in the different political systems around the globe. What is involved in this complex area of analysis can be illustrated by some relatively clear examples. Having done that it is then possible to consider the different ways in which power may be analysed and the ends of power evaluated.

Consider the problem of air pollution in the famous Italian city of Florence. From its ancient roots and its Renaissance pre-eminence through to its present, Florence has combined closely settled areas of housing and artisanal workshops with the grand monuments to its past. This has made Florence one of the great tourist centres of the world, but the number of tourists who flock to the city each year, coupled with a greater urban population, have meant an increased scope for environmental damage. Air pollution in the central area of the city is an intense and obvious problem as vast quantities of tourists, bus operators, normal cars and scooters rush their way through narrow streets, over the bridges in and through the city centre. Gradually and reluctantly, the city authorities have restricted traffic access to the centre. The problems of pollution have continued and a new campaign is under way to increase the zone subject to traffic restrictions. Some local residents and shopkeepers are openly hostile to these plans and campaign against them. Posters are stuck on many doors and in many shop windows claiming that the city (or their area) will die if the restrictions are imposed. Life will leave the city as shops close and its character will be irrevocably changed. On the other hand, the problem of air pollution is continually cited as a reason for action. Power operates on both sides of this debate. The local state, in trying to rectify a serious environmental problem, has to use power to have a regulation imposed and policed to get a change in a pattern of human and social behaviour with negative consequences for, in this case, an urban environment. The local residents are seeking forms of power that will stop this regulation being effective. Their means to power include argument, publicity, lobbying and demonstrations and may even include blockades to illustrate and press their claims. Here power to regulate and power to resist are counterposed.

Consider another kind of example. In the USA, major international mining companies have been involved in standard development versus environmental concern disputes. For example, in Wisconsin, Kennecott Copper, part of the global RTZ (Rio Tinto Zinc) enterprise, has sought to build an open-cut copper mine against the opposition of a coalition of Chippewa and white environmentalists (Gedicks 1993). The mining company has deployed considerable quantities of power, which are based on its wealth and its economic importance, to press its case and to marginalise its opponents. In pursuing its rights to business as usual in Wisconsin, as any company operating anywhere in the world would, Kennecott has used its power to mobilise investment (loans and shares), workers, technology, and political access and the ability to influence popular and elite opinion to protect its corporate goals. To succeed, the Chippewa and their environmental supporters have to find sufficient power resources to deploy to deny Kennecott Copper its normal expectations to mine and to profit from its mining. The proposal to mine is as much based on power as are the efforts of those who oppose mining and seek to have a project stopped or regulated.

When knowledge about the causes and consequences of ozone depletion reached a certain stage, it was possible for the United Nations Environment Programme and environmental activists to place concern for the consequences of the increased uses of CFCs (chlorofluorocarbons) on the political agenda. Once again, a whole array of economic, social and political forces sought resources to be used as power as the conflict over regulating the relevant industries intensified. The chemical companies concerned, in alliance with governments, clashed with rival companies and countries over whether an international protocol was needed to regulate the production and use of ozone-depleting substances (Benedict 1991). After much struggle and the invention of a plausible, commercially profitable substitute, the USA took the lead in pressing for international regulation, which, after complex negotiations and a number of summits, came into effect. Here can be seen companies making strategic calculations about the benefits that could be derived from continuing or ceasing the production of CFCs and government officials finding strategies to make more or less international agreements. And the whole process turned around power and the effective deployment of arguments and resources that could be made to produce the consequences being sought.

Now, these three examples do not exhaust all the kinds of resources that are invoked and deployed to produce power-like effects and outcomes. These are only illustrations to show that situations of environmental conflict and policy making all contain elements of power. What happens in any given environmental conflict is the result of the creation and the successful deployment of forms of power. Environmental regulation, environmental neglect, business as usual, politics as usual; all these involve the deployment and playing out of power. What needs to be considered now is how best to understand the patterns of power involved in these routine, repeated forms of environmental conflict and policy making.

The analysis of power and models of the policy process

There are a large number of different kinds of account of the character, distribution and effects of power in politics. The points of cleavage in these debates are not agreed but there is a broad division between accounts that treat power as a quantity or resource to be deployed and those that do not. Most of the literature uses the quantity/resource assumption and is treated first, but it should be kept in mind that there is a broad alternative that deals with the same matters but in quite a different analytical framework.

An initial, most useful, account of the different approaches in the dominant tradition has been given by Steven Lukes (1974) in a relatively old, very brief, accessible and influential volume, *Power: A Radical View*. In this book, Lukes argues that there are three broad ways in which power is analysed and each of these is based on finding new 'faces' of power, successively adding these to form a complete account of the character and consequences of power.

The first face of power was identified and celebrated by the American pluralists. Their position is exemplified in the works of Robert Dahl (1961, 1970) and Nelson Polsby (1963). Dahl and Polsby were caught up in the behavioural methodological revolution for the social sciences and emphasise changes in observable behaviour as the key indicator of the presence and distribution of power. Following the classic sociological argument of Max Weber about the provenance of power (Weber 1978: 53), Dahl and Polsby believe they can detect its presence when the wishes of one person or group can prevail over the wishes of others. In their approach the best situations for the study of power are those involving observable conflict between groups seeking to influence government policy making. Group A wants one thing, Group B another and the government acts. The extent to which A and B get what they want indicates both the presence of power and the pattern of its distribution between them. In simple terms: 'A has power over B to the extent that A can get B to do something that B would not otherwise do' (Dahl 1957: 203). Of course, this formula needs to be methodologically extended to encompass even the normal political processes, which are the focus of Dahl's attention.

The application of this methodology was relatively simple and remains attractive. Find a situation of policy conflict, say a dispute between the proponents of sand mining on Fraser Island (off the Queensland coast, Australia) and those opposed on environmental grounds. Establish their contrasting demands and the initial position of the relevant government. Consider the power resources brought to the conflict and the skill with which they are used. Identify the outcome in terms of a specific government decision or action. Find which group got most of what it wanted and plot the distribution of power accordingly. Having done so, it is possible to analyse the part that power played in the particular dispute and its resolution. In this case, since the sand mining of Fraser Island was stopped, environmentalists used their power resources more effectively than the mining companies and their supporters, and won.

The core of this position is based on a distinctive methodology and there is no particular reason why the methodology should have become associated with claims about the social distribution of power, but it did. The development of the pluralist approach came out of an attempt to refute the claims made about the unequal distribution of power in US society by elite theorists who were critical of the workings of US democracy. In contrast to the elite approach, which discovered concentrations of power, these academics found that there were no significant concentrations of power. Indeed power was diffused throughout society and the institutions of American democracy were such that diffused power resources would eventually get all voices listened to, and included in, the political process and in having their grievances addressed if not resolved. Much of the criticism directed at these 'pluralist' interpretations turns around not the question of methodology but the blatant unreality of this account of the distribution of power and its effects.[1]

There is an assumption in the pluralist/behavioural approach about the character of politics and the political process. It is assumed that people go about living their lives in whatever ways they choose or have chosen for them and that politics is a specialist sphere of society that is removed from the arena of their everyday interactions. People become political, choose to make issues political, enter the political process as a result of some disruption of, or disaffection with, the normal. A problem or grievance arises or there is a changed perception which renders something unacceptable that was previously acceptable. In these circumstances, individuals may seek political redress or, if the problem is shared, individuals might form groups (variously labelled interest or pressure groups) to put pressure on politicians to respond to their demands. Their ability to get what they want is a product of their power resources, the effectiveness of their organisation, the skill with which strategies and tactics are designed and power resources are deployed and, of course, the effectiveness of their opponents. Depending on their needs, people may join or form political parties and seek electoral success to address their problems. What is important here is the basic model of politics arising from grievance or interests with organisation and political action as a plausible response, seeking redress.

Lukes develops his account by tracing the way in which the methodological assumptions of pluralism are gradually undermined. For him, the next stage in the argument and the next face of power is found by adding conditions that undermine the utility of behaviourist assumptions. Here he focuses on the work of Bachrach and Baratz (1962, 1963, 1970) with their emphasis on agenda setting and a 'non-decision-making' power. For Bachrach and Baratz, power can be shown to be present when the actions taken by a group effectively prevent an issue getting onto the political agenda and marginalise the grievances of sections of the population in such a way that their needs never become political issues. In their own work, questions of race in American cities are used to illustrate this dimension of power. There is also a very effective account of contrasting city responses to air pollution that uses this approach in Matthew Crenson, *The Un-Politics of Air Pollution* (1971). Here the political conditions that allowed air pollution to be placed on the political agenda in one city were contrasted with those in another city where the issue did not emerge, even though the 'objective condition', air pollution from steel plants, was shared. In a sense, the second face of power revealed in these arguments probes further the assumptions about politics established by the pluralists. Politics is still about grievance, but here attention is paid to the role of power in preventing a grievance from being given effective political expression.

For Lukes (1974: ch. 3) this approach does not go far enough; too much of the behavioural methodology is still left in place and a further face of power needs to be added to give a complete picture. This face of power is not revealed by considering conditions of conflict and measuring who wins most from the resolutions of government, or by considering the deployment of power to shape a political agenda to accept some and marginalise other issues. Here the emphasis is on the way in which the interests[2] that people have may be denied even when those concerned remain unaware that that is what is involved. Power here is so effective that the very wants and desires of individuals are shaped by power to serve the ends of others. Into this realm the behavioural methodology cannot reach, and Lukes produces an argument by the invention of a 'counterfactual' to establish whose interests are served by an exercise of power even when there is no opposition and no observable behavioural change. It is not that the conception of politics has been broadened here. Rather, the process begun by consideration of 'non-decision making' and agenda setting is taken a stage further, to consider a kind of power that allows people's interests to be harmed without them being either aware of or able to formulate the grievance upon which overt political action would be based (Box 1.3).

John Gaventa (1980: 253), in his study of social life in an Appalachian valley, takes Lukes' broad points about the different faces of power and seeks to apply them to a situation in which quiescence has replaced rebellion in the midst of continuing difficult social and economic conditions. Gaventa concludes that 'power can and should be viewed in its multiple dimensions, and that mechanisms or processes within each are specifiable' (Box 1.4).

In constructing an analysis of power involved in environmental conflicts or environmental policy making it is possible to use any or all of these

Box 1.3

Lukes and the study of power

Lukes defines the concept of power:

'A exercises power over B when A affects B in a manner contrary to B's interests.'

For Lukes, an analyst needs to consider:

- instances of conflict
- conflicts which do not appear when a particular group's interests can be seen as harmed
- decisions that are made
- inaction and non-decisions which benefit one group but continue the harm to another
- issues and potential issues
- positions claimed by participants in conflict
- outcomes and consequences to see if interests are advanced or harmed.

Source: Adapted from Steven Lukes, *Power: A Radical View* (1974)

Box 1.4

Levels and mechanisms of power

First dimension of power

- conflicts over fundamental issues of inequality settled in the past by the defeat of those challenging inequality

Second dimension of power

- grievances continue today
- expressed in covert ways
- anticipated defeat limits activism
- cultural values reflect power imbalance and assist the powerful

Third dimension of power

- continual (re-)legitimation of inequality
- indirect mechanisms made effective by the culture of (anticipated) defeat such that even grievances may not arise in an overt way
- benefiting from power is made easy

'the total impact of a power relationship is more than the sum of its parts. Power serves to create power. Powerlessness serves to re-enforce powerlessness. Power relationships, once established, are self-sustaining' (p. 256).

Source: Adapted from *John Gavanta, Power and Powerlessness: Quiescence and Rebellion in an Appalachian Valley*, Clarendon Press, Oxford, 1980

dimensions of power and the methodological arguments associated with them. Elements of each of these faces of power may be present in or relevant to a particular case study. If, for example, you wanted to study how the US government came, during the Reagan presidency, to reduce the scope for regulation by the Environmental Protection Agency (EPA), it would be possible to use a pluralist/behaviouralist methodology to study the competing efforts of environmental and business groups to get the government to change its mind and act differently. It would be possible to use arguments about agenda shaping to see why and how power was used to construct an effective political agenda for deregulation that was able to marginalise the previous institutionalised form of environmental concern. Further, it would also be possible to extend the argument, by reference to Lukes' third face of power, to the impact that this had on the real interests of those affected by the decision to lessen the level of environmental regulation and surveillance.

All the above approaches can be treated together as part of a broad tradition in the analysis of power, covering liberal, conservative and Marxist positions. They share similar conceptions of politics as grievance/interest-based and power as a capacity and quantity that determines what happens, who wins, who gains and who loses by what is done. There is (and perhaps always has been) an alternative conception of power that is based on a different set of assumptions about the analysis of society and politics. The best contemporary expression of this can be found in Barry Hindess, *Discourses of Power* (1996), which, inspired and informed by a critique of the work of Michel Foucault, provides a sharp response to the quantity conception of power.

Back when behaviouralism and pluralism were conducting analytical battles with the elitists, an alternative view of power was proposed by the structural functionalist Talcott Parsons (1957). Parsons produces quite a different conception of power, which displaces much of the emphasis on conflict and the explanation of events in terms of different quantities of power in the hands of interested (that is, possessing interests) actors. His much-cited definition is as follows:

> Power then is generalized capacity to secure the performance of binding obligations by units in a system of collective organization when the obligations are legitimized with reference to their bearing on collective goals and where in case of recalcitrance there is a presumption of enforcement by negative situational sanctions – whatever the actual agency of that enforcement.
>
> (cited in Lukes 1974: 27–8; Hindess 1996: 34)

As others have observed, this ties power to conceptions of legitimacy and the pursuit of 'collective goals' as opposed to all other forms of action and conflict. This is not all. For Parsons, power is not just negative or coercive but, as he expresses it, a medium like money, one that enhances or increases the capacity to get things done collectively that would be difficult or not possible to achieve individually. It was on this basis that he rejected C. Wright Mills' analysis and the more general emphasis on inter-group conflict because it treated power as if it were only a 'zero-sum' game.

Parsons' approach is strongly conditioned by his broader theoretical conception of society and, although it enjoyed considerable prestige in its day, and a degree of imitation, it is not that influential now. It is also less easy to apply his insights to the analysis of either environmental conflict or policy making. Nonetheless, there are issues that can be illuminated by applying his basic position. The rise of environmental regulation can be seen as a case of an enhanced collective capacity/ability to act based upon the increased willingness of voters to treat such action as legitimate. Voter authorisation legitimated an increased capacity of government to act to define and secure collective goals over environmental policy regardless of any specific conflicts over particular issues.

Parsons is not the only author with this concept of power as being positive, in the sense that it supports some collective purpose, as opposed to power as negative, to constrain, limit or prevent other things being done. Arendt's work on violence (1970) also includes an argument about power as the product of legitimacy and processes of popular or group authorisation.

This line of reasoning, or at least elements of it, reappears in some of the work of Michel Foucault and in the commentary on the concept of governmentality developed by those influenced by his work (Burchell, Gordon and Miller 1991; Gane and Johnson 1993). Foucault produced a large number of studies combining arguments about power, knowledge and different forms of social and personal life, ranging from prisons to sexuality. He also wrote a number of books that deal more directly with questions of method and gave many interviews reflecting on his method, the political implications of his writing and on the interpretation of his work as a whole. Within these it is possible to see the outlines of an alternative conception of power.[3] Part of this alternative are versions of the argument about the performative/productive character of power, the coming together of people, their combination, producing the circumstances for more and different things to be achieved. Much of it is a reconfigured version of the rejection of power as a zero-sum game, expressed as rejection of the 'repression hypothesis' and an overemphasis on the negative side of the presence and operation of power. Also involved is Foucault's continual insistence on the connection between power and modes of resistance, which advertise themselves as a resistance to power but which are another face or part of power strategies. Foucault's work on power and the construction of subjectivity serves also to undermine the 'standard' of personal or inner autonomy universally used to establish repression and the negative impact of power. Further, the many metaphors Foucault produces to describe the operation of power focus his account on the conception of power as strategic and interactive rather than just as a clash of quantities of power with the distribution of power read back mechanically from an assessment of outcomes, the view that dominates in the power debate described above.

Foucault's lecture on governmentality (1991) has also been very influential for those trying to systematise this alternative 'Foucauldian' interpretation of power. In 1978, Foucault gave a series of lectures that included his discussion of 'governmentality'. This one lecture found its way into English via an Italian transcription of the original, given in French. It is important to note that Foucault never used the term 'governmentality' in his published works, although a few passages of the text find their way into the *History of Sexuality*, Volume I. As such it was not an important expression of Foucault's interpretation of social life. Since then the conception of 'governmentality' has been linked to Bruno Latour's notion of 'governing at a distance' and has been applied to a whole range of policy areas to produce challenging and contested interpretations (Dean 1991; Curtis 1995).

At its core, 'governmentality' emphasises the importance of knowledge and modes of calculation as well as internalising these in various subjectivities to explain what happens in particular institutions and social settings. For example, Miller and Rose (1993), in their article 'Governing Economic Life', use the concept of governmentality to show how a change in an accounting process was generalised and internalised in the modes of calculating investment as part of an effort to improve the efficiency of the British economy. Miller and Rose go to some lengths to show that the source of this change was not 'government' or the 'state' as such, but a whole range of diverse sources, including the autonomous actions of accountants acting as a profession. Policy and policy outcomes in this view are as shaped by the ways in which 'problems' are conceived and measured as by the efforts by those in conflict seeking to invent and use power resources. For the study of environmental politics, conflict and policy making, an easy illustration of the reordering of the interpretation of power comes from a consideration of the operation of environmental impact assessment. Frequently, environmental impact assessment is treated as a form of government regulation imposed on business from the outside, by the state, and resisted by business. Considering the same process through the interpretative lens of governmentality, attention would be paid to the way in which the institutionalisation of environmental impact assessment is linked to the development of new ways of knowing and calculating (and managing) the consequences of economic activity. Studies would reveal that new expertise is generated in the enterprise, that managing structures and processes are modified and that the way in which firms consider their products and production procedures could be changed. Such an interpretation could be easily grafted onto the interpretation of the Environment Impact Assessment (EIA) process given by Bartlett (1990) and Schrecker (1990), which focuses on the way in which 'ecological rationality' (Dryzek 1990, 1987) can invade and transform the 'administrative mind' (Paehlke and Torgerson 1990, 2005). More work needs to be done on the way in which this alternative interpretation of power can be applied to environmental conflict as well as environmental policy making. A significant introduction to this analysis can be found in the collection edited by Darier (1999).

Some key study terms

Environmental justice

Environmental justice has arguably always been the dominant concern of those in majority world countries, where many citizens' livelihoods depend upon their immediate natural surroundings, and who often suffer at the hands of wealthy national elites and wealthy foreign companies. Calls for *environmental justice* include: equity in the distribution of environmental goods, ills and risks; recognition of the diversity of the participants and experiences in affected communities; the protection of community capabilities and functioning; and participation in the political processes which create and manage environmental policy (Schlosberg 2004: 517–540). But at an even more basic level, the concept of justice in the context of the South can be understood as ensuring that basic needs for survival are satisfied. The needs that all people have access, for example, to shelter, clean water, food, are, in fact universal human rights, reflected in the International Human Rights Covenants (Barnett 2001).

The concept of environmental justice, or the uniting of both environmental and social justice concerns, has also become increasingly prominent in minority world countries, especially the US, since the late 1980s. In the US, the movement emerged in the wake of the civil rights movement with the recognition that race was a crucial factor in determining the quality of one's environment. In *Dumping in Dixie*, the ground-breaking work of Robert Bullard, it was argued that black Americans were more likely to live on or alongside toxic waste sites (1990: 32–7). It is now frequently found that low-income, majority world communities (despite living within minority world affluent countries) continue to bear greater health and environmental burdens, while the wealthier, white members of the population receive the bulk of the benefits (Sandweiss, 1998: 35). This situation is also sometimes called environmental injustice and environmental racism. Environmental justice activists in the US have focused on changing distributional inequity ranging from grassroots community empowerment to federal level changes.

Even today some of the most prominent environmental justice campaigns in minority world countries like the US and Australia, surround the locations of commercial hazardous waste facilities. This currently includes the search for high- and low-level radioactive waste depository locations in the US and Australia respectively, both targeted for traditional indigenous lands. The proposed Yucca Mountain Repository in Nevada, US, has been hotly contested by the Western Shoshone nation, while Australia's federal government is deciding on which of three locations near Aboriginal communities in the Northern Territory to impose a national radioactive waste dump. This follows a successful campaign by the Kupa Piti Kungka Tjuta (senior Aboriginal women of Kupa Piti) of South Australia in rejecting a waste dump on their sacred lands. The indigenous peoples of both Nevada and South Australia have already experienced the legacy of nuclear testing on their traditional lands.

See Box 1.5 for the principles adopted by the First National People of Color Environmental Leadership Summit, 1991, which have served as a defining document for the grassroots movement for environmental justice.

Box 1.5

Principles of environmental justice

Adopted at the First National People of Color Environmental Leadership Summit, 1991, Washington, DC.

Preamble

We, the people of color, gathered together at this multinational People of Color Environmental Leadership Summit, to begin to build a national and international movement of all peoples of color to fight the destruction and taking of our lands and communities, do hereby re-establish our spiritual interdependence to the sacredness of our Mother Earth; to respect and celebrate each of our cultures, languages and beliefs about the natural world and our roles in healing ourselves; to insure environmental justice; to promote economic alternatives which would contribute to the development of environmentally safe livelihoods; and, to secure our political, economic and cultural liberation that has been denied for over 500 years of colonization and oppression, resulting in the poisoning of our communities and land and the genocide of our peoples, do affirm and adopt these Principles of Environmental Justice:

1 *Environmental justice* affirms the sacredness of Mother Earth, ecological unity and the interdependence of all species, and the right to be free from ecological destruction.

2 *Environmental justice* demands that public policy be based on mutual respect and justice for all peoples, free from any form of discrimination or bias.

3 *Environmental justice* mandates the right to ethical, balanced and responsible uses of land and renewable resources in the interest of a sustainable planet for humans and other living things.

4 *Environmental justice* calls for universal protection from industrial by-products and the extraction, production and disposal of toxic/hazardous wastes and poisons that threaten the fundamental right to clean air, land, water, and food.

5 *Environmental justice* affirms the fundamental right to political, economic, cultural and environmental self-determination of all peoples.

6 *Environmental justice* demands the cessation of the production of all toxins, hazardous wastes, and radioactive materials, and that all past and current producers be held strictly accountable to the people for detoxification and the containment at the point of production.

7 *Environmental justice* demands the right to participate as equal partners at every level of decision-making including needs assessment, planning, implementation, enforcement and evaluation.

8 *Environmental justice* affirms the right of all workers to a safe and healthy work environment, without being forced to choose between an unsafe livelihood and unemployment. It also affirms the right of those who work at home to be free from environmental hazards.

9 *Environmental justice* protects the right of victims of environmental injustice to receive full compensation and reparations for damages as well as quality health care.

10 *Environmental justice* considers governmental acts of environmental injustice a violation of international law, the Universal Declaration On Human Rights, and the United Nations Convention on Genocide.

11 *Environmental justice* must recognize a special legal and natural relationship of Native Peoples to the US government through treaties, agreements, compacts, and covenants affirming sovereignty and self-determination.

12 *Environmental justice* affirms the need for urban and rural ecological policies to clean up and rebuild our cities and rural areas in balance with nature, honoring the cultural integrity of all our communities, and providing fair access for all to the full range of resources.

13 *Environmental justice* calls for the strict enforcement of principles of informed consent, and a halt to the testing of experimental reproductive and medical procedures and vaccinations on people of color.

14 *Environmental justice* opposes the destructive operations of multinational corporations.

15 *Environmental justice* opposes military occupation, repression and exploitation of lands, peoples and cultures, and other life forms.

16 *Environmental justice* calls for the education of present and future generations which emphasizes social and environmental issues, based on our experience and an appreciation of our diverse cultural perspectives.

17 *Environmental justice* requires that we, as individuals, make personal and consumer choices to consume as little of Mother Earth's resources and to produce as little waste as possible; and make the conscious decision to challenge and reprioritize our lifestyles to insure the health of the natural world for present and future generations.

Environmental security

> The environment has often been used as a tool of war, from the salting of Carthage to the Russians' scorched earth retreats before the armies of Napolean and Hitler. Plato, mocking the notion of a republic of leisure, argued that such a regime would soon resort to a war to satisfy its taste for more space and natural resources. But sustained thinking about the environment–conflict connection is a product only of the last few decades. While clashes over non-renewable resources such as oil or gold are as familiar as the Persian Gulf war, the question now is about the role of renewable resources such as water, fish, forests, and arable land.
>
> (Dabelko 1999: 14)

Definitions of environmental security are as numerous as definitions of what constitutes the 'environment' itself, and the issues involved are as diverse as biological and ecological security; the greening of military operations; climate change; desertification; biodiversity; human population and migration; fisheries; forests; energy; water; nutrition; shelter; and poverty.

The interest in environmental security emerged forcefully in the Brundtland Report in 1987, and increased at the first Earth Summit in Rio de Janeiro in 1992. The nexus between environment, development, and security was never stronger than at the 'Earth Summit Plus Ten' in Johannesburg in 2002. However, the notion of environmental security is strongly contested between those who cast the definition around the security of the nation-state, and those who use a more inclusive definition which transcends nation-state boundaries, looking at conditions which secure individual access to a basic infrastructure for survival in geopolitical regions defined by shared environmental boundaries.

The first and most common variation is concerned with the impact of environmental stress on societies, which may lead to situations of war within and between societies. In this manner, environmental security agendas are about seeking issues which, if not addressed, may provide the basis for increasing human conflict, viewing environmental stress as an additional threat to peace and stability; the securitisation of the environment by nation-states. This is a negative understanding of environmental security (see Huntington, Mayer 2004, for this kind of approach). Within this framework, when people are seen as part of an 'environmental security' agenda, ambiguously they are not perceived as part of the environment, but are simply users or, in the case of the poor, degraders. In 1990, the United Nations Human Development Report argued that poverty is one of the greatest threats to the environment, and in 1993, the International Monetary Fund (IMF) announced: 'Poverty and the environment are linked in that the poor are more likely to resort to activities that can degrade the environment' (International Monetary Fund cited in Broad, 1994). There are two key problems with this line of argument. First, all poor people are regarded in a homogeneous fashion, rather than existing in vastly different *types* of poverty with different relationships to their environments. For example, there are those still operating (threatened) subsistence lifestyles, those who have been recently removed from this lifestyle, and those people who have long ago been driven to the precipice of survival (the 'landless and rootless'). Second, many Western environmental security theorists fail to weigh up the costs of advanced industrialism on a global scale, and issues of massive overconsumption in the minority world.

A concept of environmental security which is more inclusive of the interests of the majority of people in the world, is one that moves away 'from viewing environmental stress as an additional threat within the (traditional) conflictual, statist framework, to placing environmental change at the centre of cooperative models of global security' (Dabelko and Dabelko 1995: 4). But to do this, there must be increased understanding of the environment, not as an external enemy force, but as a diverse nature which is inclusive of people; a nature which has the potential to provide secure access to individual citizens of all countries to basic nutrition (for example, protecting the *food sovereignty* of nations and communities against transnational commerce); adequate access to healthy environments; appropriate shelter; and a security to practise a diverse range of liveliehoods which are both culturally and ecologically determined (Barnett 2001, Dalby 2002, Dodds and Pippard 2005, Doyle and Risely 2008).

Environmental citizenship

Ecological citizenship has been promoted predominantly by Andrew Dobson as an example and an inflection of what he calls 'post-cosmopolitan' citizenship (Dobson 2004). There have also been other theorists attempting to define ecological citizenship and they agree that it involves a set of moral and political rights and responsibilities among humans, as well as between humans and nature. Ecological citizens should be able to pursue their own private interests but also keep in mind their obligations to the surrounding environment. For these authors, developing an ecological citizenship is considered necessary for true sustainability, as it is by considering the environment as a fundamental ethical component of citizenry, without the reliance on market-based incentives which are promoted by governments at present, that the environment will be fully integrated into the public sphere.

There have been debates over whether ecological citizenship is an original theory, given its inability to ascribe membership and its similarity to other forms of citizenship in terms of virtue. Dobson maintains that recognizing that the environment is a motivating force and central feature of ecological citizenship changes the dynamic relating to, for example, citizenry virtues and that this makes the case for a distinctive definition (Dobson 2006).

Conclusion

Environmental politics is as varied as the issues, the activists and the political systems in which it is practised. All kinds of political variables combine to condition what happens and its significance for particular environments. In terms of political regime it does make a difference if environmental movements seek to achieve their goals in liberal-democratic representative regimes or under various forms of authoritarian, military or one-party rule. It also makes a difference how the institutions of politics are designed and linked. These factors together make up the institutional setting within which environmental conflict and policy making take place.

To assess both the character and fate of environmental politics requires a knowledge of these institutional factors, their implications for environmental activism and the issues that are the basis for environmental mobilisation. More than that it requires an understanding of the different arguments about how the role of power in politics can be assessed. This chapter has outlined the key concepts and alternative arguments needed to produce an effective interpretation of environmental politics either in a single country or in a number of countries. The comparative approach always strengthens our ability to be precise about the factors that condition the outcome of environmental conflict and policy.

The chapter has concentrated on the institutional, structural and 'permissive' factors that surround environmental activism. It is now necessary to turn to the active element, the way in which different individuals and social forces respond to the environment as an issue.

Further reading

Crick, B. (1964) 'The Nature of Political Rule', in *In Defence of Politics*, Allen Lane, London.

Darier, E. (ed.) (1999) *Discourses of the Environment*, Blackwell, Oxford.

Dobson, A. (2004), *Citizenship and the Environment*, Oxford University Press, Oxford.

Doyle, T. and Risely (2008) *Crucible for Survivial: Environmental Justice and Security in the Indian Ocean Region*, Rutgers University Press, NJ, New York and London.

Hindess, B. (1996) *Discourses of Power: From Hobbes to Foucault*, Blackwell, Oxford.

Leftwich, A. (1983) *Redefining Politics: People, Resources and Power*, Methuen, London.

Lukes, S. (1974) *Power: A Radical View*, Macmillan, London.

2 Political theories and environmental conflict

- Environmental resistance, reform, and radical critique
- Traditional political theories of the right and the left
- Different interpretations of green economics
- Role of religion in environmental thought

Introduction

Having analysed the character of the political process and the nature of prevailing political regimes, it is now necessary to consider some examples of the way in which politics has been applied to the question of the environment. Depending on frameworks of understanding, different pictures or collections of ideas rise to the surface for analysis. Within this chapter we use three different axes which depict and, in turn, influence the manner in which the theoretical underpinnings of environmental politics are considered: (1) the scale of incorporation of environmental concerns from resistance to theories of reform and revolution; (2) the intersections between traditional political theories from both the right and the left, and their shaping of different notions of green economics; and, finally, the interplay between the cultural and philosophical dimensions of diverse religions and environmental theory and action.

Using these three disparate frames it is then possible to explore the whole plethora of ideas and claims that have motivated environmental activism, prompted the development of the complex of environmental movements and fuelled political and policy conflict over environmental questions. This chapter provides the foundation for a subsequent consideration of the internal character and dynamics of the various environmental movements that now operate in most countries around the globe.

Theories of environmental concern: resistance, reform and revolution

Initially it is useful to examine the ways in which some very significant groups and governments have refused to accept that environmental damage ought to be treated seriously and have resisted efforts to put environmental concerns on to the political agenda. The contrast can then be established with those governments that have been willing to incorporate degrees of environmental concern as add-ons to 'normal' politics, or have been willing to reform administrative procedures and institutional structures to bring environmental considerations inside the policy process. Finally, we will examine some green theories that demand urgent and fundamental political change.

Resisting the environment as a political issue

From the time that environmental concern started to be expressed in the United States of America and in Europe, those committed to growth were quick to mock its seriousness and relevance to politics. In the USA, figures like Herman Khan were conspicuous in their efforts to make a polemical assault on the findings and arguments of the Club of Rome (Meadows *et al.* 1972) and promulgate an alternative, optimistic futurology. Others were eager to join in and mock the seriousness of both environmental issues and those who sought to do something about them. Significant among these was John Maddox, editor of the influential British science journal *Nature*. His book, *The Doomsday Syndrome* (1972), was a very good example of the style, using his faith in science to debunk various claims about resource depletion, population growth, the negative consequences of an overemphasis on economic growth, and the increased scientific and technological redesign of agriculture.

Throughout the 1970s and 1980s, as the levels of environmental concern grew and activists struggled to get environmental issues on to the policy agenda, a continual stream of scepticism was poured on this endeavour by certain kinds of economists. All were concerned with portraying growth as the answer to economic, environmental and social problems. Some promoted free-market or market-like solutions to questions of environmental management (North 1995). Certain themes are stressed in these accounts. First, most claim to have some level of sympathy with environmental concerns but reject extremism and calls to urgent action. It has to be noted that evidence of their levels of environmental concern is hard to come by but some, like Wilfred Beckerman (1990), did try to promote the use of market instruments to deal with pollution. Even when doing so it is open to interpretation whether he was motivated more by a concern for the consequences of pollution or revulsion at the thought that government might use environmental concern to impose regulation on the economy. A similar ambiguity is to be found in most of the 'free'-market advocacy on environmental policy.[1] For example, in Bennett and Block (1991), a joint Canadian–Australian production, more time is spent trying to refute the case for environmental concern than in trying to define free-market solutions to these environmental problems. Again, the real enemy would seem to be the prospect of state intervention.

This genre of literature, focused on hostility to the consequences of environmental concern, enjoyed its prime in the early 1990s, by which time its themes had become quite predictable (Beckerman 1995; North 1995). There is no need for anxiety about population levels since human ingenuity and biotechnology will provide the food needed for whatever future levels of population there are; concerns about organochlorides are exaggerated, the product of chemiphobia, and the benefits outweigh the costs; the science of the enhanced greenhouse effect is complex and uncertain and there is no need for precipitate action (or any other kind, except research); the level of species extinction is exaggerated and the case for biodiversity is weak (besides, how many whales do we need anyway?); there is no problem with finite resources since human ingenuity, technology and price responses to increased scarcity will provide. Those who think otherwise are described as sincere but misguided extremists, zealots and Cassandras, undermining a proper and balanced concern with environmental problems and a recognition of the virtues of economic growth (North 1995).

The case for environmental scepticism wrapped up in free-market policy nostrums has not been without significant political consequences. In the United States of America, presidents over the last quarter of a century have often undermined attempts to promote green policies. In the 1980s Ronald Reagan campaigned to emasculate environmental regulation in the name of free-market, pro-business reform. As a result, his head of the Environmental Protection Agency was cited for contempt of Congress and the head of Superfund (a scheme for cleaning up contaminated waste sites) served a term in jail for trying to enact his vision of a regulation-free environmental policy. This influence lingered on, symbolised by George Bush Senior's refusal to sign key protocols of the Rio Earth Summit in 1992 and, throughout the rest of the 1990s,

Clinton's willingness to compromise a whole range of environmental measures, all in the name of economic efficiency and the global competitiveness of the US economy.

In 2001 George W. Bush came to power with a far more unilateralist and pro-business, pro-hydrocarbon approach, refusing to sign the Kyoto protocol. By 2006, however, with oil prices soaring, even this president admitted to seeing the long-term writing on the wall, suggesting in the State of the Union address that America needed to break its addiction to oil. Nevertheless, he saw the solution to this problem, not in any change of behaviour or modification of the neoliberal economic model, but through technology, including the highly contentious concept of 'clean, safe nuclear energy'. He saw the solution to the immense and diverse social and environmental problems caused by America's addiction to the car, not in providing more public transport, but by developing utopian 'pollution free' cars. In addition, although the wars embarked upon by a 'Coalition of the Willing' were rationalised by 'a war on terror', or the eradication of 'weapons of mass destruction', at least, in part, these military engagements can also be understood within frameworks of energy security; particularly oil.

Proposing reform

Politics with a green tinge

Given the success of environmental movements in having environmental issues put on to the agenda, it took only a little while for the political system to respond, even if that response was little more than just adding a tinge of green to the justification of politics as normal. In some cases, much more was done to institutionalise environmental concern than that.

The shallowest response can be seen in the statements of governments like that of Margaret Thatcher's in the late 1980s. After dramatic evidence of the consequences of pollution in the North Sea, Margaret Thatcher made a bold commitment to incorporate environmental concern into the policy making of her government. There was supposed to be a statement of national environmental goals with regular reports from government departments on how these goals were being met. A public relations flourish accompanied the announcement of these changes, but it was not long before the normal calculations of the Thatcher regime reasserted themselves. By 1996 it was recognised that the processes set up had been a waste of time as they were not being used to shape government policy at all. In this case, the only greening of politics came in the rhetoric used to describe what was already being done and to marginalise those who sought greater commitment to policy change.

Sustainable development

Dressing up existing policy as if it contained green initiatives was not the only way of reaching an accommodation with rising levels of environmental concern. In this context, and with increased tension between the developed capitalist countries and those poorer countries seeking to emulate them (symbolised by the debates over UNCTAD (United Nations Conference on Trade and Development) and the Brandt Report), a great deal of effort has gone into producing a concept of sustainable development that sought to combine the virtues of environmental concern with the pursuit of economic development and growth (Dryzek 1997: 123–36). These efforts have largely come from work sponsored by the United Nations Environment Programme and the Stockholm and Rio 'Earth' Summits. At Stockholm in 1972 the concept of sustainable development was only just being hinted at (Ward and Dubos 1972). It acquired some definition with the publication of the *World Conservation Strategy* (IUCN 1980).[2] This was followed up by some clarification and redrafting of various national conservation strategies (in Australia, the UK and the USA). All this was subsumed in slightly more detailed accounts in the Brundtland Report, *Our Common Future* (WCED 1990)[3] and the debates that flourished surrounding the Rio Earth Summit and the publication of *Agenda 21* (1992).

Although the arguments generated to link the concerns of economic development and environmental concern have varied over time (McEachern 1993), there is a core set of propositions associated with the concept of sustainable development. At its heart is the claim that the scope for sustainable economic growth is linked to the survival of sustainable environments. The flow of natural resources needed for economic production, the capacity of the soil to sustain food production, and the health of air, rivers and oceans, are all required to be able to imagine sustainable and repeatable economic growth and development. The slogan 'sustainable development is development that meets the needs of today's generation while not impairing the needs of future generations' captures much of what is at stake in the invention of the claim to shape policy by the requirements of sustainable development.

It should be noted that sustainable development is not a radical environmental or green concept, since it accepts the prime need for economic growth and the dominance of human welfare over the needs of the environment; and it conceives the relationship between humans and nature in terms of the use of the environment by and for humans. Nonetheless, adopting sustainable development does require a substantial rethinking of the terms of policy calculation and policy making and has at times been vigorously opposed by business and economists (Beckerman 1995: ch. 9), although many businesses have been willing to adopt the term as a best defence of their actions against environmental criticism. Further, the adoption of a sustainable development policy framework has sometimes been associated with political strategies to deal with the political consequences of rising (unpredictable) environmental concern.

The best example of this comes from the efforts of the Hawke Labor Government in Australia (1983–1991) to co-opt environmental concern to its electoral survival by seeking to codify the policy settings for ecologically sustainable development (ESD). Australia adopted sustainable development through a consensus-building process that concluded with the issuing of the National Conservation Strategy for Australia (1982). However, this document was not treated seriously and business adopted sustainable development to defend its existing practices from environmental criticism. Levels of politicised environmental concern intensified and the government became involved in a number of deft political manoeuvres, trading off environmental protection of specific sites for endorsement from environmental organisations in increasingly close elections. As a final twist in this saga, the government sought to incorporate environmentalists by involving them in the drawing up of new ESD policy frameworks. This time the focus was on broad policy parameters for the key sectors of the economy (such as manufacturing, mining and agriculture) and key intersectoral issues (such as urban systems and the response to the greenhouse effect). Drafting the ESD (1991) reports was the task of several working parties made up largely of bureaucrats with smaller numbers of business figures, environmentalists and trade unionists. No effort was made to find a consensus, although there seems to have been wide agreement over what should be in the recommendations. ESD turned out to be a more explicit sectoral version of sustainable development, characterised by a strong commitment to market environmental economics and certain minimal rules for assessing the environmental consequences of economic development. Rather like sustainable development, ESD was used as an indicator of the government's environmental concern but it did not interfere with its continuing promotion of economic development or the speeding up of project approvals (by short-circuiting the kinds of environmental baseline studies suggested in the ESD reports). Although ESD never became important for policy making, the reports do provide a good starting point for considering what sustainable development would look like if any government ever seriously wanted to implement it.

In more recent times, the conservative Howard Government in Australia (1996–2007) has also moved from projecting itself as always resisting environmental concern to accommodating it, continuing to use the language of sustainable development. But the forms of accommodation have changed dramatically since Labour's ESD days, allowing the Howard Government to control environmental concern within its own redefined frameworks, and then to side-line it from mainstream public debate. Through its own brand of green accommodation, the Howard Government has successfully disempowered and discredited environmental concerns in a number of ways. First, it has renamed its own environmental (green) agenda as a 'brown' one, with most of its 'environmental' monies, derived from the partial sale of Telstra (a previously state-owned telephone company) being directed towards rural, primary industries. Second, it has removed or reduced funding for its most vociferous critics like the Friends of the Earth and the Australian Conservation Foundation. In turn, additional monies and support have been directed to the politically palatable and more narrowly focused and non-domestic 'nature conservation' organisations like International Humane Society (IHS) and Worldwide Fund for Nature (WWF). Next, wherever possible, it has removed globally recognised green issues, like Greenhouse, climate and forests, from its national agenda. Moreover, it has decimated two decades of environmental legislation, and has replaced it with the Environmental Protection and Biodiversity Conservation Act (EPBC), which hands over much of the Commonwealth's determination on environmental issues to the more resource exploitation-oriented states. Critics of this Act argue that it defines environmental concern in a constrictive 'Noah's Ark' form, overly focusing on threatened species, the RAMSAR convention for protecting birds in wetlands and the word 'biodiversity'. Effectively, these critics argue, it is an agenda which is fundamentally apolitical and, therefore, one which does not challenge business and politics-as-usual. It must be said, however, that Howard's version of accommodation was made possible by Labor's retreat from environmental issues under Keating. Particularly after the election campaign of 1993, fought largely on economic issues (and the impact of the Fightback! package), Keating felt the ALP Government no longer owed a debt to the environment movement. Consequently, he would only tolerate dealing with the movement if it were 'Government-friendly'. In this vein, he commented to staffers after his 1993 election win: 'Now the environment is back where it belongs', meaning now it was a non-issue, compared with the 1987 and 1990 campaigns when the green vote proved important in the ALP's re-election. This era contrasts with the cosy relationship the peak environment groups, and especially the Australian Conservation Foundation (ACF), had with the Hawke Government during the years of ESD negotiations (Doyle 2001).

In very recent times, there are signs that the issue of climate change has re-envigorated the politics of environmental concern within parliamentary circles leading up to the 2007 Federal Election. The Howard Government remains committed to fossil-fuel and nuclear technology, whilst the ALP is positioning itself as promoting more alternative forms of energy, urging the Government to sign the Kyoto Protocol (see Chapter 6 for further discussion of climate change issue).

Radical environmental critique

There is little radicalism in the position of those who advocate either resistance or accommodation to environmental concern. Their actions combine to promote, at most, a slow pace of environmental reform with a great emphasis on broad policy statements, policy documents and the creation of 'new' institutions to look after environmental issues. Very few of these initiatives, with the possible exception of those sponsored by the United Nations Environment Programme, came from an environmental awareness that was not shaped by a response to a vigorous environmental politics generated by far more radical environmental movements. Environmentalists involved in these movements argue that green incremental change to 'business as usual' is not enough. They demand more deep-seated, widespread change in order to attain their environmental objectives.

There is a wide variety of radical environmental positions, including those on the left and the right, and those based on spiritual and religious influence, which are discussed later. There is also the radical politics of eco-feminism which crosses the full gamut of left–right green political philosophies. It is not possible, in the space available, to do much more than sketch the main features of these different arguments. There are very good surveys and arguments about green political theory that can be consulted for further details (Dobson 1995; Eckersley 1992; Goodwin 1992; Merchant, 1992, Hay 2004).

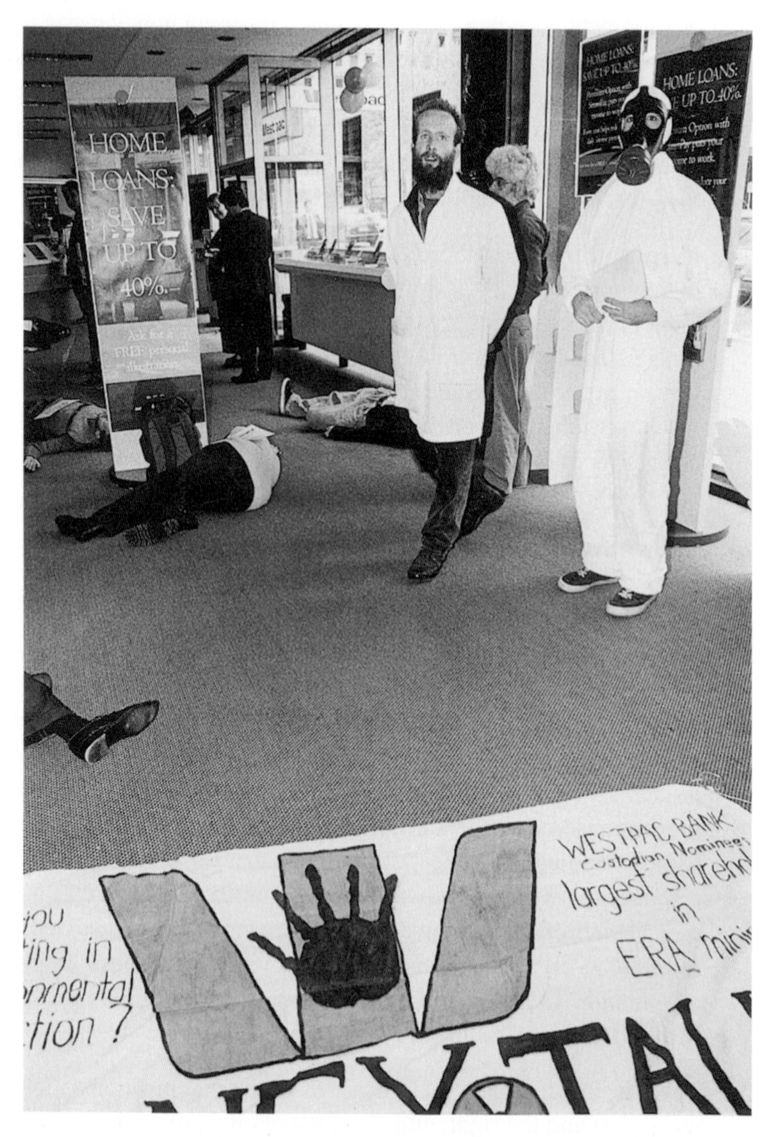

Plate 2 ***Westpac 'Die-in', Westpac national day of action.This action was part of a campaign organised by the Jabiluka Action Group against banking interests which were funding the development of the Jabiluka uranium mine in Australia's Northern Territory. Courtesy of Jabiluka Action Group, Adelaide***

Most of these eco-radical philosophies propose paradigmatic change. Thomas Kuhn, in his immensely influential work *The Structure of Scientific Revolutions* (1969), described how bodies of scientific knowledge are created. Contrary to popular belief, Kuhn argued, those ideas that combined to make up 'best science' were not necessarily the best ideas available at any one time; rather, they were the ideas that 'belonged' to the most powerful group of scientists. Kuhn referred to this constellation of ideas, beliefs and values as the dominant paradigm. This paradigm of knowledge could not accommodate new ideas sufficiently, as traditional knowledge was defended fiercely by its advocates. Consequently, revolutions in science necessarily occurred in cycles, replacing dominant paradigms with a completely new and radical constellation.

Radical environmental political theorists are involved in paradigm struggles, each seeking to create new sets of key values and principles that directly challenge existing, powerful paradigms. Each theory portrays the environment in a state of crisis due to the dominance of these powerful 'ways of seeing'. Some of these 'new values' are shared by a number of eco-radical theories, although each theory argues its position differently. For example, deep ecologists (see Table 2.1) level the bulk of their criticism at the anthropocentrism (human-centredness) of the dominant paradigm. Social ecologists (see Table 2.2) argue that hierarchy is the problem. The eco-socialists attack capitalist principles as the main culprit. Ecological post-modernists identify modernity as the paradigm that must be challenged and overthrown. Finally, certain types of eco-feminists argue that the paradigm of patriarchy is largely responsible for widespread social and environmental destruction.[4] We now turn to just two of these positions which identify different 'environmentally destructive' paradigms – modernity and patriarchy – and then challenge them.

Ecological post-modernism

The environmental movement has always included those who have rejected the whole social project of the Enlightenment. These people see environmental damage as a product of 'Enlightenment thinking', which extolled the virtues of rationality and the ability of humans to use 'nature' for their own ends in the name of progress and an ever-increasing standard of living. Technology and the whole project of modernity is seen to have produced as many problems as

Table 2.1 ***Feminism and the environment***

	Nature	*Human nature*	*Feminist critique of environmentalism*	*Image of a feminist environmentalism*
Liberal feminism	Atoms Mind/Body dualism Domination of nature	Rational agents Individualism Maximisation of self-interest	'Man and his environment' leaves out women	Women in natural resources and environmental sciences
Marxist feminism	Transformation of nature by science and technology for human use	Creation of human nature through mode of production, praxis	Critique of capitalist control of resources and accumulation of goods and profits	Socialist society will use resources for good of all men and women
Cultural feminism	Nature is spiritual and personal	Biology is basic Humans are sexual reproducing bodies	Unaware of inter-connectedness of male domination of nature and women	Woman/Nature both valorised and celebrated Reproductive freedom
Socialist feminism	Nature is material basis of life: food, clothing, shelter, energy Nature is socially and historically constructed	Human nature created through biology and praxis (sex, race, class, age)	Leaves out nature as active and responsive Leaves out women's role in reproduction and reproduction as a category	Both nature and human production are active Centrality of biological and social reproduction

Source: Adapted from Merchant (1992).

Table 2.2 ***Key differences between dominant attitudes to the environment and those of deep ecology***

Dominant attitudes	*Deep ecology*
Domination over nature	Harmony with nature
Nature a resource, intrinsic value confined to humans	Natural environment valued for biocentric egalitarianism
Ample resources or substitutes	Earth supplies limited
Material economic growth a predominant goal	Non-material goals, especially self-realisation
Consumerism	Doing with enough/recycling
Competitive lifestyle	Co-operative lifeway
Centralised/urban-centred national focus	Decentralised/bioregional/ neighbourhood focus
Power structure hierarchical	Non-hierarchical/grassroots democracy
High technology	Appropriate technology

Source: Sylvan and Bennett (1986).

it has solved and the claims of progress are fairly hollow. For these critics, the final achievement of progress and reason will be ecological destruction.

For much of the time, these views could be called anti-modernist, since they define their position in terms of a rejection of modernity. These anti-modernist views have produced some of the most enduring images of contemporary environmental protests, since it was the anti-modernists who rejected the codes and practices of affluent lifestyles and sought simpler ways of living, in tune with nature. In many places, these people left cities, sought out the countryside and the forest, and mounted their critiques of the overdevelopment of the cities.

In recent years, there has been an accommodation between these anti-modernist views and the post-modernism and post-structuralism of the academy (Gare, 1995; Conley, 1997; Cheney 1989). Post-modernism, in simple terms, sees Western societies as transformed by modernism to such an extent that modernist principles and beliefs no longer hold and, through excess, turn into their opposite. Here the link can be made: post-modernists too see the products of rearranged technological processes delivering increasingly negative consequences for the Earth and its populations. Some of these post-modernists embrace a dethroning of reason and a more fragmented, decentralised social world.

Post-modernists tend to emphasise the importance of locality and difference as against centralisation and the notion of a homogenised sameness. Post-modernists and post-structuralists would, on the whole, be sceptical of claims about an 'essential nature'. Nature itself would be seen as a social construct and the relations between humans and nature would be capable of almost infinite variety.

Many are extremely hostile to this environmentalist critique of modernity and Western science. Martin Lewis writes:

> In one sense, the romance between ecoradical philosophy and post-modernism is entirely natural. Post-modernism began its career with an attack on the sterile architecture of high modernism, and what, after all, could be more antithetical to the organic ideal of the greens than Bauhaus 'machines for living'? Similarly, the poststructuralist offensive against the supposedly disembodied, logocentric, objectivist, totalizing, imperialistic, Eurocentric project of Western science and rationality conforms impeccably with the ecoradical critique of Western philosophy. Concordance can also be found in the emphases placed on diversity in cultural expression and in nature itself. Even the postmodernists' accent on playfulness finds its echoes in the Greens' favoured account of natural processes.
>
> (1996: 219)

Much of Lewis' depiction of eco-radicalism is based on an attack on deep ecology, since deep ecologists often share this disdain for Western science with their ecological post-modernist counterparts. There are, however, major differences between them. Whereas deep ecologists see 'nature' as some transcendental force with intrinsic value, ecological post-modernists would deconstruct this depiction of nature and treat it as a social construction with invented 'value', intrinsic or otherwise, a fairly standard human pretension.

Eco-feminism

Eco-feminism emerged in the 1970s as part of the women's liberation movement. Carlassare (1994) argues that the term 'eco-feminism' is utilised 'by some activists and academics to refer to a feminism that connects ecological degradation and the oppression of women'.

The women's movement remains one of the most vibrant movements advocating social change at the turn of the twenty-first century. Like the environmental movement, feminism is full of important and different perspectives, far too many to do justice to here. Caroline Merchant identifies four positions that can usefully be discussed: liberal feminism, Marxist feminism, cultural feminism and social feminism (see Table 2.1; Merchant 1992: 186–7). She writes:

> Liberal, cultural, social and socialist feminism have all been concerned with improving the human/nature relationship and each has contributed to an ecofeminist perspective in different ways. Liberal feminism is consistent with the objectives of reform environmentalism to alter human relations with nature from within existing structures of governance through the passage of new laws and regulation. Cultural ecofeminism analyzes environmental problems from within its critique of patriarchy and offers alternatives that could liberate both women and nature.
>
> (1992: 184)

Social eco-feminism is more closely attuned to the eco-anarchist writings of Bookchin. The similarities between the two are many, with emphases on decentralisation and attacks on all forms of hierarchy, particularly patriarchy.

> Social ecofeminism advocates the liberation of women through overturning economic and social hierarchies that turn all aspects of life into a market society that today even invades the womb.
>
> (Merchant 1992: 194)

Finally, socialist eco-feminism is closely intertwined with socialist ecology (eco-socialism); but in this case, it is reproduction not production, 'which is central to the concept of a just, sustainable world' (ibid.: 195).

As there are many divisions between different radical environmental camps, so too these divisions manifest themselves within eco-feminism. One of the most interesting discussions revolves around the 'essentialist/ constructionist' tension (Carlassare 1994: 221). The essentialists, and some cultural feminists, argue that women are closer to nature than men, and this innate quality should be recognised, further developed and celebrated. Some liberal, social and socialist eco-feminists, of course, abhor this suggestion, arguing that social factors, not innate qualities, are seen as the most important shapers of gender-based inequalities (see Salleh 1992 and Cuomo 1992 for these more constructionist arguments).

Using the essentialist/constructionist axis, Lewis refers to the essentialist eco-feminists loosely as 'radical eco-feminists', while the constructionists are more 'mainstream'. He contends:

> To many feminist thinkers, the notion that female thought is fundamentally different from male thought in any way – especially by being more in tune with nature – is both offensive and dangerous. This line of reasoning . . . has long been used by men to keep women in subjugation. When women are identified with nature, men are able to appropriate the realms of science and rationality for themselves.
>
> (1992: 35; see also Plumwood 1988)

Vandana Shiva is another eco-feminist who attacks the essentialists, this time from a self-conscious Third World perspective. She sees colonialism and patriarchy as explaining the success of the globalisation of 'Western development' and advanced capitalism. She explains:

> Development was thus reduced to a continuation of the process of colonization; it became an extension of the project of wealth creation in modern Western patriarchy's economic vision, which was based on the exploitation or exclusion of women (of the West and the non-West), on the exploitation and degradation of nature, and on the exploitation and erosion of other cultures.
>
> (1994: 273)

Shiva goes on to argue that this particular kind of oppression explains why women and other subjugated cultures have been direct activists in opposing modernisation and development in parts of the Third World, 'struggling for liberation from development just as they earlier struggled for liberation from colonialism' (ibid.: 273; Mies and Shiva 1993).

Left vs. Right

As there is no one environmentalism, there is no one green economics. Different economic frameworks and solutions are built on different political theories and systems. Many of these systems depict 'natural processes' in different ways. Many of these frameworks and systems use elements of environmental determinism to justify their political and economic paths as 'right' and 'true'. Let us now briefly review the depiction of nature found in five traditional political theories taken from the much used, but little understood, left–right theoretical continuum: conservatism, liberalism, and radical libertarianism on the right; and from the left, socialism and anarchism. Advocates of all these disparate theories have developed their own forms of green economics (Doyle 2001).[5]

Green economics and 'the right'

Conservatives, liberals and radical libertarians (also referred to here as neo-liberals) are all part of 'the right'. Their advocates usually argue for green economic solutions to be found within the frameworks of both advanced industrialism and capitalism (politics is subservient to economics). Interestingly, each ideology houses profoundly alternative views on what 'human nature' is. Conservatives argue that humanity is innately evil. A wonderful example of the conservative political philosophy is found in William Golding's work *The Lord of the Flies* (Golding 1954). In this famous book a group of boys are stranded on a deserted island without the recognised authority of adults to guide them. Without this, the boys resort to 'the rule of the beast', which lies at the heart of our 'human nature', and develop a society based on greed, fear, coercion and force. One of the morals of the story is the necessity of authority imposed upon citizens 'from above'. It is not surprising then, that conservatives would support monarchies and aristocracies, for example, as they would propose that 'ordinary people' are limited in their capacity to resolve their own problems. Indeed, conservatives argue that there is a 'natural order' which *hierarchically* allocates certain privileges and powers to certain castes within societies. This *variety* is 'natural' and ensures a smoother running of a given society than if its operation were determined by such things as democracies.

'Natural history' is seen by conservative political theorists as something which is intrinsically violent: a universe at war. How then has conservatism intertwined with environmentalism? Eckersley comments:

> Now it is certainly true that there are some notable points of commonality between conservatism and many strands of environmentalism. The most significant of these are an emphasis on prudence or caution in innovation (especially with respect to technology), the desire to conserve things (old buildings, nature reserves, endangered values) to maintain continuity with the past, the use of organic political metaphors, and the rejection of totalitarianism.
>
> (Eckersley 1992: 21)

One of the most obvious instances of the conservative green economic response relates to 'scarcity economics'. The earliest proponent of these economic solutions was Thomas Malthus, 'the Dismal Parson' (Malthus 1826). He presented his 'population principle' two hundred years ago. In his principle, Malthus argued that the Earth's resources would increase arithmetically, while the Earth's population would grow geometrically: this meant that the Earth's resources were finite, and could not support an ever-expanding population.

This line of reasoning has been used to justify birth control programmes and to urge no response to poor people in need as a result of famine and drought. It was an extremely popular argument in the 1960s and 1970s, championed by both Garrett Hardin and Paul Ehrlich. Their argument begins with the assumption that there are already too many people consuming the Earth's limited resources. A 'population bomb' is set to explode. With such a narrative other questions emerge: what is the Earth's 'carrying capacity'; and who should have access to increasingly scarce resources?

The debate about 'carrying capacity' within green movements is an essentially conservative one. It assumes 'natural limits' placed on earthly existence by some form of transcendental moral authority: 'Nature' with a capital 'N'. The conservative green economic response, in this case, has largely been about reducing absolute numbers of human beings, usually from the less affluent world. Poverty *per se* is seen as one of the main causes of environmental degradation; not a result of it. 'Australians for an Ecologically Sustainable Population' is one such conservative, green organisation offering economic solutions along these lines by opposing further immigration on the basis of ecological limits. There are elements of this conservative response which appear in the radical green ideology of deep ecology (see Table 2.4), though other tenets are derived from other traditional political philosophies, such as anarchism.

The liberal tradition, though still from the right, sees natural processes, including human nature, in a substantially more generous light. Classical liberal theorists like John Locke and Jeremy Bentham argued that the vast majority of people are good; but they are sometimes hampered by an anti-social minority. So nature, in this light, is not in a perpetual state of war, but can get into minor skirmishes from time to time. In a bid to protect the majority from the anti-social minority, liberal theorists argue for the establishment of a moderate state, not an authoritarian Leviathan,[6] to promote the interests of the majority. Although liberalism does have some notion of 'greater good' it is still heavily based on the rights of the individual, attempting to allocate autonomy to each citizen within responsible bounds umpired by governments. Liberalism is heavily reliant on private property and is closely associated to the politics of capitalism with its emphasis on continuing economic growth. It is also based on pluralist assumptions that all citizens have equal access to power and resources, often denying inequities based on class, gender, race and species.

The liberal green economic response is sometimes designated as 'shallow ecology'. Most obviously, it has led to the establishment of laws and legislation which provide environmental protection, while not interfering unduly with the everyday mechanics of capitalism. One instructive example of a liberal green economic response was the development of much environmental legislation in the 1970s, and also the establishment of the Environmental Protection Authority (EPA). The EPA – both state and federal – in Australia was established to monitor the excesses of an environmentally degrading minority and to punish its excesses through a series of fines, impositions and further regulations.

Radical libertarianism (or neo-liberalism), the third ideology housed in the right, is still a form of liberalism, but is a particularly extreme, simplistic and, currently, virulent strand. Much of the radical libertarian/neo-liberal or 'market liberal' response is extremely reminiscent of laissez-faire economic systems which emerged in Europe in the last century, coinciding with the industrial revolution. These economic libertarians advocate their own interpretation of Adam Smith's proposition that within capitalism exists *an invisible hand*, whereby the pursuit of individual self-interest leads, unwittingly, to the advancement of the common good. In this way, they seek maximum autonomy for the individual, and regard any government interference as a direct threat to this freedom and autonomy. It is rather obvious what 'the state of nature' is to radical libertarians: the survival of the fittest individual leads to the survival of the fittest collective, though the *condition* of the collective is hardly important. This is remarkably similar to popular and often misguided interpretations of Charles Darwin's theories of 'natural selection' and Herbert Spencer's theories of social evolution. It is hardly surprising that both emerged on the coat-tails of the industrial revolution and the initial surge of capitalism. This most extreme version of liberalism is really an anti-society ideology. Society is nothing more than a collective of atomised individuals; it is not designed to promote infrastructures of welfare, health and education, let alone protect and uphold a collective sense of 'environmental good'. In the tradition of laissez-faire, the poor and the sick simply deserve to be so, and the 'management' of nature is best left to market forces, as the market *is* natural. It is this right-wing tradition which now dominates US and Australian politics under the conservative coalitions.

Green radical libertarian approaches have also been renamed free-market environmentalism. Attempts to *green the national accounts* and *green consumerism* provide excellent examples (Eckersley 1993: 29–30). In the first instance, green free-marketeers argue that a key problem is that not all of nature has been allocated economic value. Robyn Eckersley comments:

> Many natural resources are regarded as gifts of nature with a zero supply price, and the accumulation of wastes is regarded as a 'negative externality'. For example, our woodchips exports are registered as national income but no allowance is made for 'natural capital depreciation' (i.e. the depletion of our forests)
>
> (Eckersley 1993: 29)

So rather than arguing that there are some natural attributes which are beyond fiscal value, green radical libertarians seek to include all of nature in their market strategies[7] in a bid to resolve environmental problems whilst still pursuing economic growth under capitalism. Interestingly, while libertarians insist on the user-pays principle elsewhere, they usually reject the *polluter-pays* principle.

Green consumerism is another central plank of all liberal green responses, but it is most avidly championed by the radical libertarians. Because of liberalism's associated commitment to the role of the individual, much emphasis is placed on the environmentally educated consumer. The argument goes: 'Once people start purchasing more environmentally friendly products, then the market will have to respond and produce more of the same.' In addition, with so much emphasis on consumption, there is the notion that producers will self-regulate. The solution, however, is still firmly entrenched in the politics of pluralism and capitalism which assumes that producers exclusively respond to markets, rather than setting and influencing the marketing agenda.

Green economics and 'the left'

Leaving the various ideologies of the right, let us now focus on two traditional political ideologies with their origins derived from the left: socialism and anarchism, which, when coupled with environmentalism, become eco-socialism and social ecology.

> Eco-socialism is anthropocentric (though not in the capitalist-technocratic sense) and humanist. It rejects the bioethic and nature mystification, and any anti-humanism that these may spawn . . . Thus alienation from nature is separation from part of *ourselves*. It can be overcome by reappropriating collective control over our relationship with nature, via common ownership of the means of production: for production is at the centre of our relationship with nature even if it is not the whole of that relationship . . .
>
> Eco-socialism defines 'the environment' and environmental issues widely, to include the concerns of most people. They are urban based so their environmental problems include street violence, vehicle pollution and accidents, inner-city decay, lack of social services, loss of community and access to countryside, health and safety at work and, most important, unemployment and poverty.
>
> (Pepper 1993: 232–4, italics in original)

The eco-socialists (or socialist ecologists/environmentalists) attempt to mesh ecological principles with those of another, more traditional set of political theories revolving around Marxism. In this instance, its ecotopian visions very closely match those of certain types of socialism. Socialist ecologists are anthropocentric enough to believe that they should move strategically from issues of social justice to ecology, not vice versa.

There are many forms of socialism and communism and this discussion gives only a broad overview. Socialism, as a political tradition, pre-dates its reformulation by Marx and Engels in the nineteenth century, but it was the Marxist version that was most influential, having inspired the state-centred versions of both the USSR and China, as well as the radical left critics of social democracy in the liberal democracies. Socialism is focused on both social inequality and the consequences of a class-divided capitalist society. In their famous pamphlet *The Communist Manifesto*, Marx and Engels claimed:

> The history of all hitherto existing society is the history of class struggle.
>
> (1848)

Marx's writings were an elaborate and, at times, savage critique of capitalism. He saw capitalism as creating two great, opposing classes in society: the bourgeoisie and the proletariat. The bourgeoisie owned the basic 'means of production' such as factories, machinery and the income-producing property. The proletariat did not, and was totally dependent on the only commodity it possessed, its ability to work, or in Marx's analytical vocabulary, its 'labour power'. As a consequence of their different relationship to the means of production, the class of labour had to sell their labour power to capital. This allowed production and consumption to occur but defined the relations between the classes in terms of a fundamental antagonism. On the basis of this structured opposition and the experience of exploitation, class conflict would be continuously generated. Marx saw the conflict between classes as the 'motor of history'. Inevitably, capitalist development would produce 'wealth at one pole, misery at the other', and an increasingly polarised society. It was Marx's expectation that sharp class antagonism would provide the basis for radical political change when the proletariat united to overthrow the social, economic and political power of the bourgeoisie and usher in a new stage in world history, socialism followed by communism.

Socialists do not view nature as capitalism often portrays it: as an amalgam of atomised individual wins and losses, designed to promote the fittest within individual species along the path of evolutionary victory and, at the same time, throwing the losers out onto the genetic scrap-heap. Instead, nature under socialism can be seen as a community, working for the ultimate survival of that whole ecosystem in question. The ecological concept of 'holism' nicely matches with this political thought.

This political philosophy generates its own brand of green economic response. In a more centralised green economy, the state would not only monitor and legislate against the excesses of environmentally degrading companies, but the state would actually drive a green economy by being the major shareholder in it. The issue here is not whether there are inadequate resources to go around but that, under capitalism, these resources have been wrongly concentrated in the hands of the few. Under socialism, the people, as a collective, would actually own their resources. Consequently, green socialists do not advocate the existence of population 'carrying capacity' or 'resources scarcity': poverty, then, is seen as a result of capitalism, not the fault of poor people producing too many children.

On this basis, Marx developed his own distinctive theory of population. Marx rejected Malthus' simple claim that the rate of population growth would inevitably outstrip society's capacity to provide all but misery and famine. For Marx, the population question was a sub-set of the relationship between expanding production capacities, capital accumulation and the ability of labour to defend its position. In some circumstances, the increase in population was a strategic response to the conditions of capital accumulation, where the expansion of production was dependent on a greater use of labour power. In other circumstances, the 'increase of labour power itself' could mean something other than just an increase in the size of the working population. Marx rejected the reactionary, and pessimistic, views of Malthus and criticised their crude restatement of certain class positions. He was more optimistic about the development of productive forces being able to provide and alter the dynamic of population growth. It is for this reason that Marx is seen as being too sanguine about the prospects of unlimited growth without population or resource constraints; hence, resource security was treated as a consequence of bourgeois exploitation and capitalist overconsumption rather than of natural limits coupled with overpopulation.

Marx's anti-Malthusian sentiments are similar to an anarchist (social ecology) position, but in this case, hierarchies of class, gender and race are replaced by classes defined purely by their position in the processes of commodity production. (See Table 2.2 for a comparison of anarchism and socialism.)

Environmentalist critics of this socialist position argue that this inability to accept 'natural limits' must mean that Marx's writings cannot be reconciled with an ecocentric perspective. Bell comments:

> Exhortation that socialism must ultimately come to grips with the question of ecological constraints on growth is one thing, but in general socialists have certainly not seized on this issue. The dominant socialist position would still be that a socialist transition, by overturning capitalism's inefficiencies and waste, could spur the growth process: outdoing capitalism at its own game.
>
> (1986: 11)

Some scholars argue that too much emphasis has been placed on this point since Marx did not overtly object to the notion that human population numbers and natural resource limits are linked. He did object, however, to the conservative values smuggled into the Malthusian 'principle of population', which 'dictated that the axe of subsistence should fall on the necks of the poor rather than the rich' (Doyle and Kellow 1995: 45).

Whatever the nuances, perhaps this is one socialist principle that has been modified in the development of an eco-socialist position. In his defining work on eco-socialism, David Pepper writes:

> The eco-socialist response to resource questions is not merely to fix on distribution, as commentators like Eckersley (an ecocentrist) suggest.

> It says that there are no ahistorical limits of immediate significance to human growth as *socialist* development. But there are ultimate natural constraints which form the boundaries of human transformational power.
>
> (1993: 233, italics in original)

It should be noted that eco-socialism and more structuralist/Marxist understandings of environmental degradation continue to influentially inform environmental movements in many less affluent countries of the South. For example in the Philippines, the goal of many Filipino environmental networks and organisations (such as KALIKASAN – People's Network for the Environment) is not to end environmentally degrading development and resource extraction *per se*, but rather to wrest control of such development and resources from the hands of large transnational corporations (often owned by wealthy Northern elites) and deliver it into the hands of local people. It is hoped that they will then pursue appropriate and "genuine" development, with fewer negative environmental consequences (Doyle 2005: 53–4). This is not to say that the structuralist/Marxist model that frames issues as a matter of elites vs. the masses is the only theoretical influence on environmental movements in the South, but that it is an influential understanding of power which often makes revolutionary environmental action a more valid prescription than in the post-industrialist, post-materialist nations of the North.

The final political theory we will visit here is anarchism. Whereas conservatives are pessimistic in their 'view of nature', and liberals are rather more hopeful, anarchists are the eternal optimists. Anarchist theorists see nature as an inherently beautiful place. They argue that, in its spontaneous form, nature is innately good, as are its people. Authority is not developed through necessity to protect people from themselves; but rather as a means for the powerful to promote their own interests, and to subjugate the interests of others. The anarchist tradition is hostile to the state, liberals and Marxists, stressing instead the importance of non-hierarchical human social organisation, decentralisation and self-government. In an anarchist society, environmental problems would be solved with maximum local autonomy, as society is seen as connected rather than severed from the ecosystem within which it resides. Each decentralised local community would be strongly connected to a specific 'bioregion', and would maximise individual autonomy within the collective whole. This brand of eco-anarchism is known as 'Social ecology', and is a position taken by a small but highly influential group within the broad spectrum of environmental movements.

Social ecologists argue that the marriage between ecology and anarchism is more than just one of convenience. They argue that there are fundamental similarities between the theories as they both identify some evolutionary telos that will deliver humanity and 'other nature' to some higher plain of existence. The fashionable catchwords of ecology normally reserved for describing 'healthy' non-human nature – interconnectedness, diversity, symbiosis, stability, flexibility and organicism – are seen as equally applicable to the anarchist vision of human societies (see Table 2.3).

Social ecologists are opposed to all forms of domination, both human and non-human. They advocate consistency between ends and means (i.e. political methods are just as important as their end result) and they are strong defenders of the non-institutional, grassroots politics of social movements (the details of which are outlined in Chapter 3). Murray Bookchin is identified as the public face of social ecology or eco-anarchism, and like the work of many radical eco-political theorists, there is a notion of 'Ecotopia' in Bookchin's musings. He imagines a society where the principles of ecology and anarchism are one:

> If the foregoing attempts to mesh ecological and anarchist principles are ever achieved in practice, social life would yield a sensitive development of human and natural diversity, falling together into a well-balanced, harmonious unity . . . Freed from an oppressive routine, from paralysing repressions and insecurities, from the burdens of toil and false needs, from the trammels of authority and irrational compulsion, the individual would finally be in a position, for the first time in history, to fully realize his potentialities as a member of the human community and the natural world.
>
> (1980: 187–94)

Table 2.3 ***Principles of social ecology/eco-anarchism***

1 Bypass/abolish modern nation-state. Confer maximum political/economic autonomy on decentralised local communities.

2 Centrality of political philosophy of anarchism. Close connections to ecology. Both draw inspiration from each other.

3 Opposed to all forms of human/non-human domination.

4 Strong defenders of grassroots/extra-parliamentary activities.

5 Consistency between ends and means.

There are two key positions of Bookchin that need to be further considered: his theses on social hierarchy and evolutionary stewardship. Bookchin maintains that it is hierarchy in human societies that generates all forms of domination within human societies and between humans and non-human nature. Hierarchy, according to Bookchin, is a social construction, and has no 'natural' basis or essence.

> The domination of nature first arose within *society* as part of its institutionalization into gerontocracies that placed the young in varying degrees of servitude to the old and in patriarchies that placed women in varying degrees of servitude to men – not in any endeavour to 'control' nature or natural forces.
>
> (ibid.: 32, quoted in Eckersley 1992: 148)

Too often inequities within human societies are justified on the basis that this unequal stratification exists 'in nature'. For example, conservative and patriarchal aristocratic systems have been justified using the example of the lion pride, with the most powerful male being classified as king. But it is just as easy to project another model of politics on to the lion pride. Feminist ecologists often view the dominant male in the lion community as nothing more than a 'toy boy' who is used for seven or eight months for breeding purposes, then traded in for a new one. The point that Bookchin makes is that these hierarchies are human constructs which are projected on to 'other' nature to justify differences in power as somehow naturally essential within human societies. In this way, Bookchin contends, human agency is denied in its bid to shape its polity, and hierarchical divisions based on gender, class and race remain ordained as 'naturally occurring'. He further suggests that once all forms of hierarchy and domination are removed from human societies, then the separation between humans and other parts of nature will also dissipate.

Bookchin's evolutionary stewardship thesis maintains that although humans are part of nature (unlike the traditional resource exploitation/conservation position), they occupy a rather more exalted place in the evolutionary scheme (unlike a deep ecological 'equality of all species' view – see Table 2.4). Indeed, he argues that humans are 'nature rendered self-conscious' (Eckersley 1992: 155).

Like socialists, most anarchists believe an ecologically viable society is incompatible with capitalism and its need for continually expanding markets and the built-in obsolescence of consumer goods. However, rather than wait for the downfall of the capitalist system, they seek to build more sustainable economic systems from below. One example of

Table 2.4 ***Eco-socialism: some socialist-anarchist differences***

Socialism	*Anarchism*
Social injustice, environmental degradation caused by class exploitation	Social injustice, environmental degradation caused by hierarchical power relations
Class is defined by economic criteria	Class is also defined by non-economic criteria (race, sex)
Abolish capitalism first and the centralised state will wither away, because capitalism creates the state	Abolish the state first, as an independent act from abolishing capitalism, because the state creates capitalism
The state is the representative and defender of the bourgeoisie	The state represents its own interests, independently of other economic classes
Participation in conventional politics is permissible in the path to revolution	No participation in conventional politics is permissible
Revolution by subverting and confronting capitalism – experimental communities, etc. are naive and utopian	Revolution through by-passing capitalism and creating 'prefigurations' of the desired society, such as alternative communities and economies
Emphasise strength of collective political action	Tend to emphasise personal-is-political maxim and individual lifestyle reform
Working class will be major actors in social change	New social movements and community groups will be major actors in social change
Modernist politics	Tendency to 'postmodernist' politics
Need for a planned economy	Communes should self-organise, within limits, because spontaneity is important
Limited support for decentralisation	Decentralisation is vital
Anthropocentric (but not in the same way as capitalism-technocentrism)	Advocates (in social ecology) neither anthropocentrism nor biocentrism
Conceives of nature as socially constructed	Tends to see nature as external to society but the latter should conform to nature's laws and regard nature as a template

Source: Adapted from Pepper (1993).

the green anarchist economic response is fairly obvious. With its emphasis on decentralisation, Local Exchange Trading Systems (LETS) have emerged around the world, beginning in Canada in 1983. LETS is a non-profit organisation which provides an alternative currency for people trading goods and services within their local community. LETS Association Inc. describes their operation as follows:

> LETS members are a collection of people who have all agreed to buy and sell with the group in return for local currency. When a transaction is to take place, the people involved negotiate the value of the deal . . . The member who receives the goods or services authorises units to be transferred from his/her account to the account of the supplying member . . . Local currency is created by the community and can only be used in that local area. It reflects the real wealth – the people's initiative, skills and efforts
>
> (LETS South Australia 1997)

Bioregionalism is another result of the green anarchist response. Bioregions are naturally occurring boundaries which define ecological and social communities. The Murray–Darling Basin scheme for example, which crosses the political boundaries of three states in Australia, gives an excellent example of a bioregional management system adopted by both state and federal governments. Bioregionalists also argue for the importance of *place*. The concept of place sees individual humans as necessarily having both a physical and spiritual attachment to both a community of other people, as well as to a bio-physical, defined space. They reject the new-found extreme mobility of advanced industrialised societies. In this manner, green anarchist economics rejects the globalisation of economic systems with its attendant reliance on free markets and capital flight.[8]

This emphasis on the importance of place and local autonomy has led to a burgeoning of the community garden movement in some first world cultures. It has underpinned the movement by citizens' groups to reclaim common space for food production and community events, enabling communities to become more self-reliant, better able to preserve and grow cultural foods, more rooted in the place they live, while reducing the global impact of their 'food miles'.

Yet for many communities around the world, particularly in majority world countries and the people of First Nations, this is simply the way things have always been done. But also, it is important to understand that the ways in which Third World cultures have traditionally produced food and organised their economic realities within decentralised polities is now also sometimes under threat in this era of globalised production. Baviskar writes of the fight to maintain decentralised economic systems controlled by local people against the Indian state:

> The slogan of 'Our rule in our villages' calls for non-cooperation with the state, the Gandhian method of passive resistance against exploitative authority. Jan andolan (people's movement), or decentralized and non-violent collective action, is posited as a political alternative to the dominant political system . . . village self-government is theoretically consistent as a form of decentralised political action that tries to create a political alternative to mainstream politics
>
> (Baviskar 1995: 224–5)

Back in the first world, the trend towards local community electricity generation (such as in Denmark and the Netherlands where 50 per cent of energy generation is now decentralised, and Britain's May 2006 budget allocation of £50 million to support community-based 'micro-generation' initiatives) (Catchlove 2006) and the use of alternative technology, have also arguably been inspired by social ecology theory and its call for emancipatory, collectively owned technology. Other grassroots community developments such as city eco-village housing projects, community environment parks and certain greener city initiatives (see Box 2.1), are also modelled along social ecology lines.

Role of religion

The world is not exclusively a Christian one, despite its dominance in the First World. Since the 1990s, it has been interesting to note the ever-increasing interplay between various traditional religions, including Christianity, with political philosophies espousing environmental concern.

The role of religion is a significant element in determining a society's view of nature and the environment. The environment, in the more affluent North, has been dominated by issues such as 'wilderness' (mainly forests). The term 'wilderness' relies on a concept of 'true nature', without the imprimatur of humans stamped upon it: the Christian garden before the great sin. In Judeo-Christian societies, the human/nature split has been a defining feature of its cosmology. Consequently, many movements in the North bearing the green cloak may be dominated by questions such as which particular parts of nature must be hermetically

Box 2.1

Centre for Alternative Technology, Wales

The Centre for Alternative Technology (CAT) is a world-famous demonstration site for energy efficiency and renewable energy generation, generating 90 per cent of its energy requirements renewably. It's also renowned for its environmental building, organic growing, and alternative sewerage systems. It began as a community committed to eco-friendly principles and the testing of new ideas and technologies, and has now become an Eco-Centre dedicated to showing practical local solutions to twenty-first-century environmental problems. It was founded in 1973 on the site of a disused slate quarry near Machynlleth in Mid Wales, and eventually opened a Visitor Centre in 1975 to display interactive public exhibitions and generate interest in alternative technology. It is now one of Europe's leading Eco-Centres, which is run as a workers' cooperative of 28 members, and receives around 80,000 visitors every year who come to learn about their work (Ward, 2004: 97). CAT practically addresses every aspect of the average lifestyle, and ideas are promoted through displays, a free information service, residential courses, and books. CAT's mission is 'the search for globally sustainable, whole and ecologically sound technologies and ways of life' (http://www.cat.org.uk).

CAT inspired a similar development in Melbourne, Australia – the Centre for Education and Research in Environmental Strategies (CERES community environment park). It is a place which exists to initiate and support environmental sustainability, social equity, cultural richness and community participation. This is achieved through functional demonstrations, displays, education programmes, community gardens, an EcoHouse, markets, festivals and walking trails (http://www.ceres.org.au).

These centres apply the anarchist concepts of community: face-to-face democracy, humanistic, liberatory technology, and decentralisation, to overcome huge environmental challenges. They work to empower local communities to become more self-reliant, ecologically and socially sustainable, and culturally rich. In short, to become 'social ecologies'.

Source: Adapted from Ward, C. (2004). *Anarchism: A very Short Introduction* (New York, Oxford University Press)

sealed from human access; or, more profitably, how can this external environment be 'managed' more efficiently.

This view is in direct contrast to many societies in the less affluent South such as, for example, in India, which has millennia of cultural development and is dominated by Hinduism, an ancient and complex polytheistic religion. Hinduism and its antecedents permeate all aspects of society and this has had a significant impact on the world view of Indians in relation to their environment. According to Tagore, Indian society has emerged from a 'forest society' and has always seen humans as part of nature, rather than separate. He writes:

> The culture of the forest has fuelled the culture of Indian society, The culture that has arisen from the forest has been influenced by the diverse processes from species to species, from season to season, in sight and sound and smell. The unifying principle of life in diversity, of democratic pluralism, thus became the principle of Indian civilisation
>
> (quoted by Shiva 1992: 196)

Obviously, other religions, such as Hinduism, Islam and Jainism, as well as non-religious ideologies have been layered, palimpsest-like over this 'original' cosmological script; so the origins of this human inclusiveness in nature in modern India are confused by the overlayering of other types of boundaries of identity. Nevertheless, Hinduism played an important role in the protests by the Narmada Bachao Andolan (NBA) over the Narmada Dam by providing the religious underpinning for the non-violent actions of 'satyagraha' (Doyle 2005). In 1994, for example, at a 'dharma' every protester's hands were tied to symbolise their commitment to non-violence. This did not stop the protesters from being badly beaten by police, but won for the movement a deep respect from many parts of the Indian populace.

'Satyagraha' is literally translated as 'firmness in truth' and was popularised by Gandhi in the fight for Indian independence. While there are times in Hindu cosmology where violence is undertaken, such as by the *Kshatriya* warrior class, Gandhi's non-violent interpretation of Hinduism gained currency and permeated all levels of Indian society. Satyagraha is more than just a case of heightened risk: it is a term which embraces an entire 'way of living'; rather than just being a 'passive resistance tactic' at a time of conflict. At the personal level it may relate to vegetarianism and acts of self-discipline. At the societal level, satyagraha is a dramatically radical political philosophy which strikes at the very heart of the modern nation-state.

The philosophy and practice of satyagraha, in organisations like the NBA, led to the education, empowerment and mobilisation of Indian villagers in the Narmada Valley and therefore played a crucial role in the relationship between religion, politics and the environment (Doyle 2004).

A similar movement that uses a non-violent philosophy derived from its religion is that of 'engaged Buddhism' which employs Buddhist philosophy to challenge Western-style capitalist development at the expense of nature. Engaged Buddhists see the rapacious development associated with capitalism and consumerism as at odds with the underlying tenets of Buddhism, which emphasises human-centred development but only when in harmony with nature. As an example, 'Right Livelihood' is considered one element of the Noble Eightfold Path of Buddhism, but it would be difficult to achieve Right Livelihood by being employed as a lumberjack, butcher or other 'violent' occupation. There have been various approaches undertaken throughout Asia by engaged Buddhists to either promote more Buddhist-friendly development or prevent environmentally destructive projects.

In Sri Lanka the long-running Sarvodaya Shramadana Movement, set up by Dr A.T. Ariyaratne in 1958, involves at least 11,000 villages in its rural development programme which aims at building a Buddhist-inspired 'no affluence, no poverty' society where concern for individuals is linked to a concern for nature and appropriate small-scale development.

In Ladakh on the Tibetan plateau, the Ladakh Ecological Development Group has introduced a farm guest programme where non-Ladakhis can spend time on a Ladakhi farm learning traditional Buddhist farming techniques. These techniques take into account the fragile ecosystem of the plateau and, at the same time, develop a harmonious symbiotic relationship between the Buddhist farmer and their environment. By introducing outsiders to these techniques, the group hopes to preserve the knowledge of the fast-disappearing traditional farming systems by raising awareness of the philosophy behind them and raising their status in the eyes of locals and foreigners alike.

In Thailand engaged Buddhists have undertaken both educational and political activities in promoting a Buddhist view of nature and society. The Spirit in Education Movement (SEM) runs courses such as 'Voluntary Simplicity for the Middle Classes', which attempts to put Thailand's new affluent class in touch with their Buddhist heritage. At the same time 'forest monks' concerned by rampant deforestation in the country have undertaken tree-ordaining ceremonies in order to protect trees from being felled. In a devoutly Buddhist country such as Thailand this attempt to mix religion with environmental politics has gained currency over the years and has now become a regular feature of environmental disputes (Simpson 1999).

The relationship between Islam and the environment is also increasing in visibility. Islam is a monotheistic religion with the same roots as Christianity and Judaism, but in the West the religion is often painted as associated with extremism and violence. As a counterpoint to this perspective the first green Islamic group in the UK, the London Islamic Network for the Environment, aims to increase understanding of the true philosophical underpinnings of Islam and in particular, its relationship to the environment. As with the Buddhist SEM it runs educational activities that link core Islamic features with green issues including seminars with topics such as 'Ramadan and Consumerism'.

Similarly, in the affluent global North, there are numerous Christian environmental organisations that link concern for the environment with Christian values. While there are some theorists who criticise Christianity for its 'man's domination of nature' or less absolutist 'stewardship' approach, some Christians see their role as earth's protectors. One such group in the US is Target Earth which, in its own words is,

> a [national] movement of individuals, churches, college fellowships and Christian ministries motivated by the biblical call to be faithful stewards of everything God created – to love our neighbors as ourselves and to care for the earth
>
> (Target Earth, n.d.)

The movement sees its role as Christians as both 'serving the earth [and] serving the poor' and conflates its missionary role with protecting nature. While some Christians may not consider these two roles as complementary – the Christian right in the US still pursues the 'domination of nature' or 'stewardship' ideology – Target Earth harmonises these sometimes conflictual pursuits engaging in 'buying up endangered lands, protecting people, saving the jaguar, sharing the love of Jesus, feeding the hungry, and reforesting ravaged terrain' in fifteen countries. At times Christian spirituality, like many religions, has also been linked with nationalism and tradition and invoked in support of environmental issues (see Box 2.2).

There are also environment movements or groups that combine more than one religious viewpoint. In the less-affluent global South, for instance, there are numerous cases of Islamic and Christian groups working together for a common environmental cause. An important example here can be found in the Philippines where adherents of those two religions have joined forces to oppose exploitative mining in the

Box 2.2

Dispute over Zengö Hill, Hungary

In 1994 the Hungarian parliament approved the building of a NATO security radar on Zengö Hill in southern Hungary. After a huge anti-radar campaign involving locals, activists from the region, environmental activists from Budapest's green underground subculture, pacifists (anti-NATO groups), international activists and networks, conservative activists opposing the government, and media, the government finally decided in November 2005 to relocate the military enterprise 16 km away at Mount Tubes.

The reason the campaign was so successful was that there were a large number of different concerns, multiple actors, and a variety of different tactics. The proposal was seen to potentially threaten human health, lead to increased European militarism, damage a nature protection area which hosted 90 per cent of the known Banatian peony flower population, and destroy Hungarian tradition and mythology due to the dozens of myths and legends connected to the peak of the hill (even the name Zengö stems from 'the special sound of the mountain that can be heard when the wind is blowing').

It is this last element which is interesting as it led to the use of both Christian and other spiritual invocations in an attempt to save the hill. This 'sweet romantic stream', mostly influenced by the Christian traditions of the inhabitants of surrounding villages, led to the revival during the environmental campaign of numerous spiritual customs connected with Zengö. For example, there was a revival of the Easter Zengö pilgrimage where locals climbed to a cross on the peak at Easter, and in 2004 the pilgrimage was accompanied by the priest of a nearby village who ceremonially blessed the Hungarian national flag symbolising that 'we are not giving the mountain'. As well as being a biblical symbol, the pilgrimage was also seen as part of ancient Hungarian culture. The Christian frame intersected with the nationalist frame informing the movement, but there was also a spiritual dimension to the element of nature protection. Many activists saw the peak as a symbolic counterpoint to the polluted city and a place where one could be raised above the world and leave their problems behind, and had a sense of spiritual connectedness to the hill and its fragile Banatian peony flower.

Source: Kerenyi, S. and Szabó, M. (2006) 'Transnational Influences on Patterns of Mobilization within Environmental Movements in Hungary', *Environmental Politics*, 15, 5: 803–20.

country, particularly in the southern island of Mindanao. One NGO, the Philippines Churches for People's Resistance (PCPR), provides organisational support for 'liberation theologists' from all backgrounds who are engaged in environmental protest.

In these situations religious adherents often find that, at their foundations, the religions have more in common with each other when engaged in protecting both people and the Earth. In this way religions are often deinstitutionalised, removing the staid power structures that usually govern them, and returning to their emancipatory roots.

While not adhering to a religion *per se*, many within the Western environmental movement are motivated by spiritual connection or transpersonal doctrines advocating spiritual revolution. Unlike eco-socialism or social ecology which argue that repressive structures must be removed before the human spirit can fully flourish, transpersonalism advocates the necessity of fundamental changes in human spirituality *before* reorganisation can occur. A variety of these spiritual doctrines, in part, inform the radical theory of 'Deep ecology'. In many ways, deep ecology can be understood as a Westernised, composite theory which brings together certain principles derived from both eastern and Western sensibilities.

Deep ecology was a term first coined by Arne Naess in the early 1970s (Naess and Rothenburg 1989; Hay 2004). Since then, it has also been referred to as 'ecocentrism' (Eckersley 1992). The most powerful assumptions of deep ecology are fourfold. First, ecocentrists argue that all beings, humans and non-humans, possess intrinsic value. The assumption that nature is a 'resource' is 'an essential and hitherto unquestioned axiom of Western history and the economic and technological systems woven into that history' (Hay and Haward 1988: 437–8). Consequently this intrinsic value argument radically challenges this prevailing notion.

> The impulse to defend the existential rights of wilderness in precedence over human-use rights has led to a spirited challenge to the most fundamental tenet of Western civilisation, the belief that rights are strictly human categories, and that no countervailing *principle* exists to bar humanity from behaving in any way it deems fit towards the non-human world.
>
> (ibid.)

The second major ecocentric assumption is that all beings are of equal value: 'that there are no "higher" and "lower" life forms in Nature' (Matthews 1988: 10). Third, there is the central principle of interconnectedness. Eckersley explains:

> According to this picture of reality, the world is an intrinsically dynamic, interconnected web of relations in which there are no absolute discrete entities and no absolute dividing lines between the

> living and the nonliving, the animate and the inanimate, or the human and the nonhuman.
>
> (1992: 49)

The concept of the interconnectedness of all living things is a premise of many eastern philosophies, such as the writings of Tao Te Ching. A contemporary Taoist scholar, Po-Keung Ip, emphasises that in the Taoist perspective, 'man and other beings, animated or otherwise, are ontologically as well as axiologically equal' (Po-Keung Ip, 1978: 335–43). However, the origins of this perspective are not eastern alone; similar views on the interconnectedness of all parts of our existence are espoused within Paganism. Pagan groups still operate within environmental movements for this very reason (Doyle and Kellow 1995).

Finally, ecocentrists often argue that the Earth is finite in its carrying capacity, and that there are too many people on the planet: 'The flourishing of human life and cultures is compatible with a substantial decrease of the human population. The flourishing of non-human life requires such a decrease' (Sessions and Naess 1983). Deep ecologists are sometimes enthusiastic in their support of population control programmes. To deep ecologists then, human beings are not the unambiguous end or sole purpose of evolutionary progress but just another species existing on planet Earth (see Table 2.4).

While deep ecology is in the same vein as other Northern environmental traditions which are dominated by issues of wilderness and resource preservation, the deep ecology argument goes beyond just an aesthetic or religious appreciation for wilderness. It is far more philosophically radical in arguing that non-human nature has an intrinsic value that is not dependent on acts of human valuation.

Conclusion

With the exception of Germany and Sweden in the most recent period, no government has tried seriously to take environmental concern into the process of policy making. Specific environmental disputes and the occasional environmental issue may have prompted government action (such as the banning of CFCs, the setting of 'climate targets', the saving of particular areas of forest, the creation of game reserves or the regulation of pollution); but mostly whatever has been done has been done to accommodate rising levels of environmental concern and to protect the demands of economic development and economic growth.

The way in which the accommodation between growth and concern has been constructed has varied from the cosmetic to the opportunist, with rare attempts to develop and embed rules that cover the best way to manage the impact of economic development on environments. With strategies of incorporation and accommodation to the fore, there has been plenty of scope for more comprehensive environmental critiques to develop, and more radical forms of environmental politics to arise, as environmentalists see their concerns marginalised or ignored.

Further reading

Barry, J. (1999) *Rethinking Green Politics*, Sage Publications, London.

Doherty B. (2000) *Green Ideas and Action*, Routledge, London and New York.

Hay, P.R. (2004) *Main Currents in Western Environmental Thought*, University of New South Wales Press, Sydney.

Kuhn, T.S. (1969) *The Structure of Scientific Revolutions*, University of Chicago Press, Chicago.

McEachern, D. (1993) 'Environmental Policy in Australia 1981–1991: A Form of Corporatism?', *Australian Journal of Public Administration*, 52, 2: 173–86.

Pearce, D. (ed.) (1991) *Blueprint 2: Greening the World Economy*, Earthscan Publications, London.

Pepper, D. (1996) *Modern Environmentalism*, Routledge, London and New York.

3 Environmental politics in social movements

- **Social movement theories**
- **Different environmental movements: post-materialism, post-industrialism, post-colonialism**
- **'North' and 'South'**
- **Transnational environmental movements**

Introduction

Environmentalism, in all its forms, was born in environmental movements. There are many theories about what makes up a social movement, and some of these are outlined in this chapter. At the outset, what needs to be understood is that social movements are largely non-institutional. They occupy a political terrain that is often quite separate from more established institutionalised political forms such as pressure groups, parties, and the administrative and parliamentary systems of the state. It was within these non-institutional, more informal realms of society and its politics that environmental movements emerged. It is safe to say that without the environmental movements there would be little or no 'greening' of government and corporations.

Until the relatively recent upsurge of literature written on new social movements (NSMs), comparatively little has been written about this more informal realm of politics. Traditional political science has largely ignored the politics of everyday life, doubting that it is important enough to warrant analysis. Social movement theorists, however, believe that most new and transforming ideas begin life in non-institutional politics. This creative 'politics of the people' is evident in dynamic, amorphous networks, associations, grassroots groups and alliances (Doyle 2000). Rarely is this dimension governed by formal laws and statutes of association, such as constitutions.

The next chapter also includes an investigation of non-governmental organisations (NGOs) in environmental politics. While still treated as non-institutional politics in this work, some NGOs cross the non-institutional/institutional divide. These NGOs have legitimised themselves through the adoption of constitutions, setting rules of conduct and defining organisational goals. They are, in this sense, formal political institutions. To a degree, the adoption of a constitution symbolises that they are willing to work within established rules and social norms. Such NGOs are as formal as non-institutional politics gets. Considerable tensions surround the relationships of NGOs to the non-institutionalised grassroots groups and networks of social movements on the one hand, and the institutions of the state on the other. For this work, NGOs are treated as being a constituent part of social movements, along with other sub-groupings such as networks and informal groups, rather than existing as an entirely separate phenomenon.

What are social movements?

Before it is possible to analyse the dynamic and diverse character of environmental movements, it is necessary to establish some of the general characteristics of social movements. 'Social movements' is a term used to refer to the form in which new combinations of people inject themselves into politics and challenge dominant ideas and a given constellation of power. The nineteenth-century labour movement is a good example. Here people who found themselves confronting harsh industrial working conditions joined together in a myriad of small organisations or combinations to press for changes, from their bosses and from the state. In Britain, for example, such 'combination' and 'oath taking' was illegal; workers did not have the right to organise or the right to vote, and defence of the rights of private property was of great importance to the state. Nonetheless, workers struggled for a vision of a better future, achieved their basic goals, transformed the rules upon which the system was run and ended up incorporated into the changed social order. The labour movement did not achieve its radical goals, or the overthrow of capitalism, but it did change the system to the extent that it and its concerns went from being excluded to being included.

In contemporary parlance, this labour movement would be seen as an example of an old social movement in contrast to the 'new' social movements, the latter including both the women's movement and the environmental movement. These are new in the sense that they challenge a new set of dominant ideas and another constellation of power. There are new issues and concerns to be injected into the political process. Like the preceding social movements they have a radical edge and visions of a world transformed by their demands. Their radicalism is heightened by their awareness of what happened before: that radical movements ended up being incorporated and their issues and passions tamed. New social movements are not fixed identities with stable memberships and ideologies, but are constituted through ongoing debate and interaction. Their structural form often overrides barriers and borders like class, religion, established political parties and even families, to form tremendously broad coalitions of support for a period of specific struggle. Doherty (2002, 2006) has identified four elements characteristic of new social movements. Participants hold a common identity which is not simply based on ideas; they use extensive network ties in which groups and individuals take common action or exchange ideas; they are involved in public protest which can be combined with counter-cultural lifestyles; and they challenge some feature of dominant cultural codes or social and political values, going beyond mere policy change. It has certainly been the intention of these movements to disrupt the taken-for-granted routines of normal politics and to push other considerations to the fore. In this vein, not all environmental movements have these emancipatory characteristics of a social movement.

The new social movements often take up themes left over from models constructed in earlier eras and give them new emphasis and meaning. Such was the case with the re-invigoration of feminist politics, the peace movement and anti-nuclear campaigns. In some ways a revised environmental politics was the same. Environmental campaigns had existed before, even in the early stages of industrialisation, and regulations had been imposed to limit pollution, largely on health grounds. From the 1970s onwards revitalised environmental movements began working through a whole array of local, national, and international networks, groups and organisations to press their claims. Often there were direct actions, such as blockades, marches and rallies, as well as quieter attempts to lobby for policy changes and new initiatives. It is the whole array of these activities that is the subject of this and the next two chapters.

Why do social movements happen?

There are many theories about the origins and character of social movements. There is now a recognised split between what has become known as the 'American' and 'European' approaches (Morris and Herring 1987; Klandermans and Tarrow 1988; Neidhardt and Rucht 1990; Diani 1992, Doherty and Doyle 2006). Although there are important exceptions in both categories, this division can be usefully employed here. The American theories of collective behaviour, resource mobilisation, and those which relate to political process, are largely action and actor centred, while the European theorists are more 'structural' in their accounts (Diani 1992: 3–4).

Although there are important and fundamental differences between the American schools of thought, many of their advocates also insist on a shared teleological dimension, amongst other things, tying participants together; defining the movement (Doyle 2000). The most significant of these have their roots in the sociological theories of the Chicago School (Princen and Finger 1994: 48). Princen and Finger write:

> Social movement theory goes back to psychosociology and the study of individual behaviour within groups. Collective action, according to this theory, can be triggered in various ways, depending essentially upon the theoretical framework to which one refers. One can distinguish three main schools. All of them are fundamentally ahistorical. Indeed, collective action can occur either as a result of relative deprivation, as a strategy to articulate common interests, or as a response to economic or political conflicts. In a political context the purpose of collective action is social change.
>
> (ibid.: 48–9)

In collective action theories, individuals are treated as rationally responding to forms of deprivation or to some newly presented opportunity for political success. Such situations usually occur when there is 'rapid social change' (Oberschall 1993: 18). In these circumstances traditional relationships and ideas are challenged, sometimes giving rise to social movements. Oberschall writes:

> In this view, a period of rapid social change – due to industrial growth and economic transformation, urban growth and rural decline, an economic depression, the aftermath of a lost war, rapid population growth, and the like – will weaken and undermine stable groups and communities . . . As social bonds weaken and traditional answers and remedies no longer work, the population will manifest signs of

> increasing disorganization . . . they participate in major social, political, and religious movements that seek to reform and restructure institutions.
>
> (ibid.: 18)

Basically, these collective action theorists believe that something must go substantially 'wrong' for people to coalesce into new social movements.

Collective action theories tend to be very general and ahistorical in their accounts of the development of either old or new social movements.

The more European tradition, which is often referred to as the New Social Movement (NSM) approach, has not placed such an emphasis on commonality of purpose, though this still sometimes emerges (see Touraine 1981). Instead, these theorists stress the importance of networks as the defining factor of this modern social phenomenon. Donati writes of this phenomenon in the context of Italian mass movements:

> Collective identity can only be formed through concrete and significant interaction between individuals. Its bargaining and its formation are carried out through pre-existing ties and networks, through everyday relationships and collective identities which are always present in the social system and which come to be changed and reshaped through the bargaining process.
>
> (Donati 1984: 837–59)

These more recent, European (and often Italian: Melucci, Donati, Della Porta and Diani) theories place less emphasis on the goal orientation of movements, but rather prefer to portray them as a clattering of multi-directional, three-dimensional informal networks. What is interesting about these theories is that they argue that the existence of these networks is a necessary pre-condition to the formation of symbolic identity, rather than the networks coalescing around a pre-existing series of rational goals. In considering new social movements, it is important to pay attention to the circumstances in which they arise and to the specific characteristics of both the participants and their goals.

Environmental movements as social movements

There are three broad frameworks which are used here to inform, and to make sense of, different environmental movement experiences: post-materialism; post-industrialism; post-colonialism.

One of the most significant and pervasive accounts of the origins of the new social movements has emphasised a 'value shift' in society explained in terms of a post-materialism thesis. Environmental movements, for example, are seen as possessing post-materialist values that directly contest, in a paradigmatic battle, the dominant materialist values of modern society. This argument is commonly identified with the writings of Inglehart (1977, 1990; see also Papadakis 1993). Strongly premised on Maslow's 'hierarchy of needs' (1954), the post-materialist argument is that having largely fulfilled the more basic needs of safety and security, parts of advanced industrial society are able to pursue the 'higher', more luxuriant causes, such as love and a sense of belonging, beyond the old politics of material existence. Inglehart states:

> A process of intergenerational value change is gradually transforming the politics and cultural norms of advanced industrial societies. A shift from Materialist to Postmaterialist value priorities has brought new political issues to the center of the stage and provided much of the impetus for new political movements . . . from giving top priority to physical sustenance and safety toward heavier emphasis on belonging, self-expression, and the quality of life.
>
> (1990: 66)

It is accepted that some environmental movements do seek post-materialist values and express their politics in these terms. In parts of the more affluent world, arguments relating to the aesthetic values of nature, non-human rights, the spirituality of place, and an emphasis on holism and ecology would seem to fit the post-materialist hypothesis. It should be noted, however, that not all First World environmental movements are either predicated on or seeking post-materialist values. In addition, in poorer parts of the world environmental movements can be effectively based on those old survival/security needs in situations made worse by extreme environmental degradation. The struggles of the Ogoni people in Nigeria against pollution caused by Shell are a case in point. So, whatever the strengths and weaknesses of the post-materialist thesis, it can only explain a little about the origins and character of some environmental movements.

An alternative account of the origins of environmental movements is based on the thesis of post-industrialism.[1] This position argues that advanced industrialism, championed by both the market systems of latter-day capitalism and the state-centred models of Soviet-style socialism, has pushed the Earth, its habitats and its species (including humans) to the brink of extinction. This industrial/development paradigm has promoted economic growth at all costs. Initially this pursuit of growth was rooted deeply in the Enlightenment project of the scientific and industrial revolutions, the pursuit of progress and improved living standards for all. The environment, and nature, was presented as a cornucopia of unlimited resources and abundance. The environmental costs of growth were either not recognised or treated as incidentals in the gaining of a greater good.

In more recent times, industrialism has become global and there is widespread (if partial) acceptance of natural constraints to growth, but the Enlightenment project continues. It still advocates increased growth but now this should be bolstered by improvements in environmental efficiency and management, the promotion of the global 'free market', and the advocacy of homogeneous 'democratic', pluralist political systems. Carl Boggs writes from the post-industrialist perspective:

> To the degree that the radicalism of new social movements tends to flow from the deep crisis of industrial society, its roots are generally indigenous and organic, making it naturally resistant to totalistic ideologies that galvanized the Second and Third Internationals . . . the eclipse of the industrial growth model, the threat of nuclear catastrophe, bureaucratization, destruction of natural habitat, social anomie – cannot be expected to disappear simply through the good intentions of political leaders.
>
> (1986: 23)

For writers like Boggs, the post-industrial setting generates a unique social and political climate that promotes the formation of new social movements (NSMs). As a result, the defining characteristics of these movements differ from those that went before in terms of class, and ideological and organisational characteristics.

Boggs argues that NSMs are less likely to be co-opted than movements existing prior to the 1970s, although it is difficult to see why this is more than a hope on his part. Post-industrialist theorists, however, argue that current problems are so profound that they cannot be routinised into normal politics. It is true that some parts of green movements retain their opposition to dominant institutions. But it is also true that other parts have been readily co-opted and many co-operate with government as a way of being politically effective. As with the old social movements, incorporation into a new *status quo* is likely to be the product of external struggles both within environmental movements and between these movements, business and the state.

Using post-colonialism as the narrative frame, green concerns are cast in the light of the coloniser versus the colonised; the dichotomous world of affluence and poverty; the haves and the have-nots. The NSM form may not always be appropriate for depicting the movement in the post-colonial South, being a post-positivist construction more adept at understanding the post-material and post-industrial worlds of the affluent. Sometimes more traditional models of social movements (not *new* social movements at all) based on clear sectional interests, such as labour, class or caste, may be more appropriate for comprehending the still strict divisions between those who have control over modes of production and the resources which feed them, and those that do not. Traditional social movement models based on Marxist – and most particularly structuralist – accounts of power enjoy enormous currency in the South. Large numbers of environmental activists in the developing world identify themselves as Marxists, seeing the key cause of environmental degradation being that resources and production are in the hands of a class-based elite. Solving these problems does not lie necessarily in better management or more efficient and sustainable practices. Rather, the first part of the answer lies in local peoples gaining control over their own resources, their own lives (Doherty and Doyle 2006). This movement has far from transcended class structure and economic grievances, given that national and international elites have benefited at the expense of the poor.

The authors use the terms North/South, First World/Third Worlds, developed/developing worlds, and minority/majority worlds interchangeably to describe post-colonial green politics. Most of these dualistic divisions are oriented around poverty and development issues. We acknowledge that all these terms are imperfect categories. It is a problem to define entire hemispheres as being rich and poor (Doherty and Doyle 2006). There is huge variance in levels of poverty in countries classified as part of the South. Sometimes, the World Bank uses the term 'Fourth World' to differentiate between the poorest nations and the simply poor nations of the Third World. On other occasions, it has taken out the oil-rich nations of the Middle East from its 'South' categorisation. There are also classification problems when considering recently industrialising countries versus those who have yet to undergo significant industrialisation. In some ways, one can follow the advice of the Calverts in their text devoted to discussions of the environment and North-South and simply say that the South is 'taken to mean all the countries of the world not defined as Advanced Industrial Countries (AICs)' (Calvert and Calvert 1999: 6). The problem with this approach, as Calvert *et al.* accept, is that enormous discrepancies of wealth exist within nation-states. In the Australian aboriginal situation, for example, with indigenous peoples living a fourth world existence within a first world nation-state, it becomes obvious that the South can exist within the North. Of course, the opposite is also true: elites in the South can enjoy wealthy lives akin to what is generically expected in the North. Another example is found in the United States: the forest preservation movements of the North West of the United States most obviously comprise a minority world environmental movement; whilst the US environmental justice movement, born in communities of colour, originally in deliberate juxtaposition to what it perceived as the 'white elite' environmental movement, can be usefully designated as a majority world experience. In any study of social movements, terminology must be employed which is not wholly based on a discourse relating to nation-states for, as aforementioned, social movements often traverse nation-state boundaries.

Of course, this is an overly dualistic and simplistic, but useful, broad-brush technique of highlighting differences. It is reminiscent of Guha and Martinez-Alier's construction of the 'environmentalism of the poor' (Martinez-Aliez 2004) as distinct from that of the wealthy. On the one hand, the environmentalism of the minority world is constructed as largely post-materialist: more interested in the rights of 'other nature', which are implicit in conservation, threatened species and wilderness campaigns. Contrarily, the issues we have described above, far from being post-materialist, are issues of survival. Elsewhere, Doyle (2005) uses this simple dichotomy of the minority world and the majority world. One criticism levelled at this work, as well as that of Guha and Martinez-Aliez, is that it unfairly and inaccurately represents green concerns in many parts of the industrialised world.

General theories of social movements, old and new, and the different categories of principles and meaning which give rise to them, should be treated as tools, rather than models into which all experience can be forced. In some situations, they can prove extremely useful, in others, they are inappropriate. There are many specific, diverse and contradictory factors that explain the rise of different environmental movements. There are no overriding, agreed, common goals that join all the different movements together. There is no unifying teleological purpose that drives them. There is no single causal reality that made them. Interpretations of the origins and significance of environmental movements are as contested as the movements themselves.

Organisation and structure in environmental movements

Environmental movements vary greatly in both their general and specific objectives as well as in their internal structures and modes of organisation. It is important to consider how these movements hold together internally and the extent to which they coalesce to form more coherent organisations.

When looking at the sprawling activities of environmental movements, political scientists have attempted to use, among other things, pluralist interest group models. But environmental movements are not just large interest groups or organisations. They are far more complex, are often more diverse, highly informal, amorphous in their structures, and constantly undergoing substantial redefinition (Doyle 1986; Doyle and Kellow 1995). Because interest group models treat collective political action as being driven by shared goals (that is, it is assumed what needs to be analysed), this has inhibited our understanding of the often fascinating mechanics and structures of the more informal relationships that are characteristic of a mass movement.

A vast array of informal groups, formal organisations, networks and individuals is involved in each environmental movement. This fragmentation is a reflection of a broad range of differing political ideals and policy goals, and consequent means for achieving them. It also reveals the segmented, diffuse and amorphous nature of the movement's structure. For Pakulski:

> Structures include patterns of links between movement specific groups and organisations, as well as groups and organisations drawn into the orbit of movement activities, but formed independently of them (e.g. political parties, religious bodies, ethnic organisations).
> (1991: 32)

For this very reason, it is not possible to ascertain the exact number of environmental groups operating at any one time. The movement is in a constant state of flux. As issues appear on the political agenda, groups often form. As the issue in question disappears from public view, the group may fade away also. Membership is fluid. Each movement comprises many individuals who are not necessarily 'card-carrying' members of specific environmental organisations. Hence, the overall membership, when defined by different individuals or groups, varies accordingly.

Plate 3 ***The Kupa Piti Kungka Tjuta traditional women elders stand victorious on Arcoona Station, South Australia, one of the proposed sites for the Australian Federal Government's national waste repository. Courtesy of Joel Catchlove***

Due to their fragmented nature, a study of the structure of specific movements is an extremely complicated one. There are five different structural forms that dominate movement activity (see Figure 3.1). Each one of these forms has distinct features, and a given environmental movement is made up of the sum of these different structures. The term 'palimpsest' has been utilised to produce a visual representation of this complex set of combinations (Doyle 1991: 3; Doyle and Kellow 1995).[2] Doyle and Kellow write:

> The primary reason for using this word from the Greek is that it has a semantic definition which is useful in the visual presentation of the proposed model. A palimpsest is a parchment from which writing had been imperfectly erased to make room for another text. The net result of this practice is a document with several manuscripts still visible, mapped unevenly onto each other.
>
> (ibid.: 90)

The appropriateness of this analogy seems most apparent in terms of the structure of environmental movements, with their three-dimensional space and different levels of political activity found and labelled within them. Individuals are linked by interconnecting lines on the diagram.

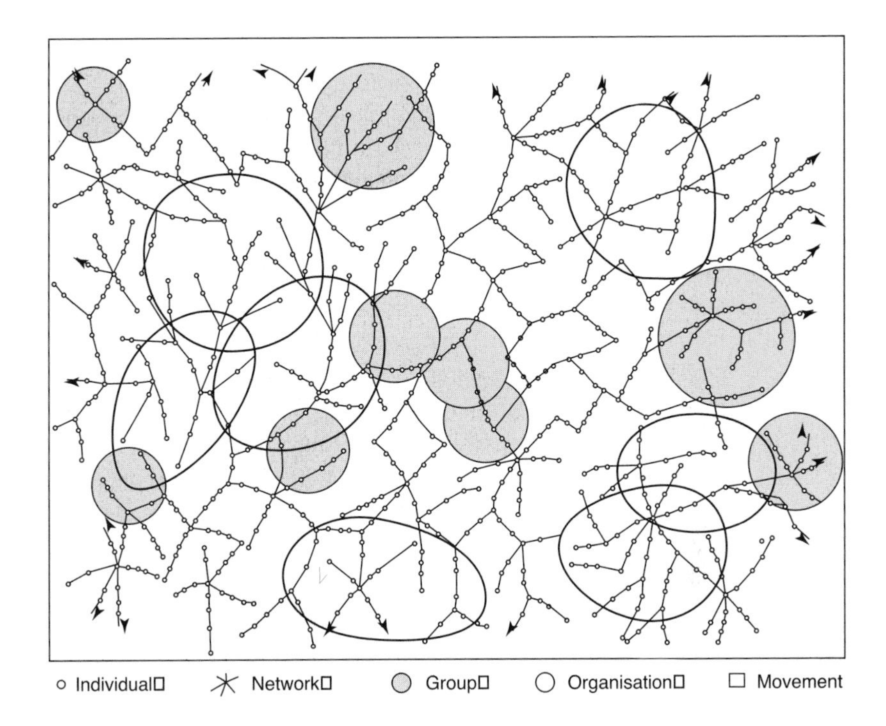

Figure 3.1 ***The palimpsest (Doyle 1991; Doyle and Kellow 1995)***

These links depict many, varied networks of individuals. Each network is different as its definition relies on the perceptions, biases and power plays of the initiator(s). The key differential variable of the network, however, is the common goal(s) or ideology that bind(s) the participants together. In most cases these goals or shared values are specifically issue-oriented, whether based on an environmental campaign or a particular type of political system.

These interconnecting lines ignore organisational and group boundaries. For these networks are fundamentally concerned with relationships between individuals operating inside and outside other formal and informal collective forms. Consequently, on occasions, these networks do not intersect with any formal organisational or informal group members. Instead, these links may indicate information exchange and other informal activity. It is these networks that provide the cement which both directly and indirectly links the different parts of the movements together.

Positioned between the levels of the network and the organisation is the *group*. The 'group' shares certain similarities with the network. The group, however, is more permanent than the network; its relationships have become more solidified over time. As a symbol of the more stable status of the group, each group has a perennial title. The creation of a collective, invariant title that symbolises the politics of the individuals and differentiates them from others is a unifying characteristic of groups.[3]

The group also differs from the organisation, which is the fourth level of the palimpsest. It is portrayed on the diagram as a solid, continuous line, symbolising the rigidity and the permanency of its well-defined, often hierarchical structure. These organisations (or NGOs) are the focus of the next chapter and are a component part of social movements. Pivotal to this concept is a vision of a movement existing as some sort of collection of concentric circles, with the organisations (or NGOs) occupying the centre, and all others inhabiting the outer periphery. This is a rather insubstantial perception of what a social movement is, and it illustrates the biases of the organisational sociologist at work. Instead, Piven and Cloward have a far more acceptable, though less defined, differentiation between movements and formal organisations. They state:

> The stress on conscious intentions in these usages reflects a confusion in the literature between the mass movement on the one hand, and the formalised organisations which tend to emerge on the crest of the movement on the other hand – two intertwined but distinct phenomena.
> (1979: 5)

The final, and most recently applied, 'manuscript' is at the level of the movement. Each movement is represented on the palimpsest as a square frame. This frame symbolises the boundaries set by each individual. These boundaries are continually changing with time, as are the different forms of individual, group, organisational and network perceptual boundaries.

Environmental movements, therefore, are overarching political forms that include actors (sometimes including social movements) advocating all of the diverse eco-philosophies and positions (and many more) touched upon in the previous chapter.

Varieties of environmental movement: post-material, post-industrial, post-colonial

Despite the incredible variety of environmental movements operating in countries around the globe, it is still possible to produce a description of the different types of movement and their most common geopolitical location and prominence. Such an account cannot be all-inclusive. It is necessary to select the most characteristic forms to use as examples. This means, of course, that there will be a serious degree of simplification and that examples to contradict the generalisation will easily be found. The purpose here is to give a broad-brush account to help locate the character and actions of any particular movement.

North American environmental movements: an example of post-materialism

Due to the largely non-institutional nature of social and environmental movements, agreements on points of history are more difficult to find than in the *monumental* histories of formal institutions. Exact dates of origin; seminal figures; and divisions into epochs are more hotly contested than usual. This is hardly surprising when one considers the amorphous nature of environmental movements and their subjective definitional boundaries. Despite this, most American writers date the emergence of the movement to late in the nineteenth century. This early movement shares certain traits with similar *nature conservation* movements which emerged in Europe, Australia and most parts of the minority world at roughly the same time.

The reading of more recent events which characterise modern environmental movements is more heavily debated. For example, Lester sees two major periods emerge within the modern American movement: 1960 – 1990; and then from 1990 until the present. The thirty years after 1960 were characterised as one which integrated the conservation and preservationist concerns of earlier movements 'with a broader set of ecosystem concerns – such as air and water quality, threatened and endangered species, pesticides, and protection of prime farmland' (Lester 1998: 3). Since 1990, Lester argues that the environmental movement has broadened its constituency even more. New sets of issues have been added to the environmental agenda: global warming, biological diversity, and sustainable development.

In a more detailed exposition of the movement in the United States, Mark Dowie divides this modern period into the same number of categories or 'waves', but they begin at different times and the categorisation is based on different factors (Dowie 1995a). The 'second wave' (the first being in the late nineteenth century until the early twentieth) began with 'Earth Day' which was celebrated all over the United States in 1970, and this wave 'ushered in the era of protective legislation in the early 1970s and was abruptly ended by Ronald Reagan in the 1980s' (Dowie 1995a: 35–6). It was at this time that mainstream environmental NGOs began their move from being outsiders to insiders, but this transition most truly occurred during the 'third wave' under President Bill Clinton.

Both writers also refer to a current/future wave: in Lester's case a 'fifth' wave, and in Dowie's, a 'fourth'. This recent wave is far more grassroots oriented; is intensely democratic in its decision making; is less middle class; is more militant, and operates outside of established bureaucracies. Importantly, this wave reconnects with radical elements of past American environmental movements, but it also challenges its elitism. This most recent collection of networks is often referred to as the *environmental justice movement* (see Box 1.5 for attributes).

Whereas most significant Western European environmental movements have been rooted in anti-nuclear energy issues and the post-industrialist politics of human and political ecology, environmental movements in the United States, Canada, New Zealand, Australia and some parts of Scandinavia have been dominated by post-materialist, wilderness-oriented perspectives with an emphasis on the 'importance of place'. These are not unlike Western European nature conservation movements but with less emphasis on the preservation of the built environment. Whereas the political ecologists of Western Europe have advocated wide-ranging social change along the lines of the four pillars outlined by the German Greens (participatory democracy, social equity, non-violence and ecology) wilderness-oriented movements have been dominated by a concern with ecology. Part of the reason for this is obvious. These countries still have large tracts of relatively 'undeveloped wilderness' (Eckersley 1990: 70) that can be saved, and a suitably large portion of the population is rich enough to define environmental concern in this way.

What is the environmental movement in the United States like? Obviously, at different times, in different places, the movement has had a variety of different concerns. On the east coast, with such a high population density, it has been impossible to escape from the human element in ecology. Consequently, air- and water-quality issues are rated highly. Unlike the political and human ecology movements, radical, systemic political change is rarely proposed by the movement, dominated as it is by powerful non-governmental organisations (the subject of the next chapter). These NGOs construe nature in a rather limited, instrumental fashion, not unlike the view of government bureaucrats and the corporate think-tanks with which they often work. The 'environmental crisis' is not really seen as a crisis at all but as a challenge for better management. Environmental problems, in this sense, are seen as efficiency optimisation projects played out in the marketplace.

The overriding focus in the western regions of the United States (and this dominates the national agenda) is on wilderness issues (See Cove-Mallard campaign, Box 3.1). Ecocentric arguments abound in the United States: forest wilderness issues; the reintroduction of mega-fauna (like wolves, grizzly bears and elk); and the 'management' of national parks. With so much emphasis on non-human nature (from both instrumental and intrinsic value positions) there has been little recognition of the fact that humanity is also part of nature. The major forms of political participation are through direct lobbying of government officials, and working with corporations to enhance environmental performance.

Networks of social ecologists (described in Chapter 2) provide a counterpoint to the more powerful deep ecologists and resource conservationists within the movement, promoting a radical human dimension. The social ecologists of the United States, like the eco-socialists of Europe (though coming from distinct non-Marxist philosophical positions) are raising issues over resource consumption and maldistribution. No longer is conserving nature seen as sufficient. It is now a question of who has access to these conserved resources and on what terms. More recent times have seen the emergence of an environmental justice movement (see Chapter 1). Environmentalism was seen as the exclusive privilege and domain of white people, who were more intent on preserving mega-fauna in wilderness parks than fighting for the basic living needs of the underprivileged black and coloured populations.[4] This part of the movement, more than any other, has dragged other parts of the US environmental movement into a political place where the social ramifications of being green have to be confronted.

Western European environmental movements: post-material and post-industrial movements

Some of the most dramatic, well-publicised and much discussed environmental movements are to be found in Western Europe. There are two major kinds (Rohrschneider 1991). First, there is the very traditional nature conservation movement, as mirrored in the post-materialist movements of North America and Australia. This is the oldest branch of the environmental movement in Western Europe, and during the

Box 3.1

Case study: Cove/Mallard forest campaign, US

Cove/Mallard was an eight-year campaign beginning in 1992 which brought together a diverse network of activist organisations such as Earth First!, the Native Forest Network, the Ancient Forest Bus Brigade, the Environmental Rangers, the Wilderness Society and the Sierra Club, to prevent the largest intact forest ecosystem in the US (the Greater Salmon-Selway) being opened up for roads and logging. This coniferous forest found in Idaho's 'Big Wild' was the last great wilderness in the US and home to many species on endangered lists such as eagles, lynx, wolves, wolverines, moose, mountain lions, and many types of fish.

Mainstream environmental NGOs were involved in the campaign, but it became clear to direct action environmentalists that they would only achieve limited gains through dealing with the legal system and these eNGOS. The Cove/Mallard Coalition led direct action campaigns each summer, ranging from *ecomilitia* tactics to more passive forms of resistance. An Earth First! rendezvous in the region in 1996 attempted to delay logging and gain national media coverage through activists chaining themselves to gates, trucks and equipment; pouring sand in gas tanks; draining oil from police cars; and treesits. This ongoing pressure led to President Clinton instructing the Forest Service to create an administrative rule to protect these areas in 1999, which was later scrapped by the Bush Administration.

While the environmental movement in the US is more apolitical than other places, the Cove/Mallard Coalition was more political in that it explicitly regarded the current political system as contributing to environmental degradation and, consequently, actively involved themselves in civil disobedience against that system. However, many of the acts of disobedience were relatively uncoordinated exploits of individuals, with little counter-ideology or systemic critique forming part of their baggage. The ideology of American pluralism is so often all-consuming that it is impossible to imagine different systems of human organisation. People who challenge the system are in a small minority, acting deep in forests, mostly outside of *normal* political arenas. Ultimately much of the focus of these ecoactivists is gaining mainstream media coverage for their *ecotheatre*, which is classic pluralist appeals-to-elites strategy. Having said this, it is reasonable to conclude that without the direct action politics of the forest activists within the US Northwest, logging would have continued unabated.

Source: Adapted from Doyle, T. (2005) *Environmental Movements in Majority and Minority Worlds*, New Brunwick, New Jersey and London, Rutgers University Press

nineteenth century was mainly motivated to preserve visible signs of human history in a rapidly modernising world. Other traditional thrusts included bird preservation. More recently, the contemporary nature conservation movement is motivated by pollution problems.

Second, there are the more post-industrialist, political ecology and anti-nuclear movements, which combined post-materialism with a broader New Left-derived analysis of power and developed a significantly different green ideology. They wrestled with questions of structural change and multiple forms of social inequality (gender, race, sexuality, class and bureaucracy) from their inception. These latter movements were post-industrial in that they critiqued modern industrial society from a variety of angles, questioning the abilities of modern science and technology; bureaucratic political structures; and economic systems based on 'growth-at-all-costs' to deliver people to 'the promised land' through progress. The Enlightenment goals of industrial societies, it was argued, were thought to be issuing the Earth with a death warrant.

The political ecology movement emerged at the end of the 1960s in reaction to problems of industrialised societies. There was a strong urban focus (see Box 3.2), as well as attempts to protect relatively untouched 'natural' areas. By the mid-1970s, environmental organisations usually controlled their operations. As Western European traditional parties were unwilling to incorporate their needs into the governmental process, they formed their own political parties (see Chapter 5). Also in the 1970s, with close ties to other new social movements, it developed the 'new environmental paradigm' view, which included, apart from ecological concerns over economic growth, a critique of science and technology, and a preference for participatory politics. Although there was an interest in the welfare of other species, there was a strong anthropocentric 'human ecological approach', with a 'nation-state' political focus.

Although characterised as a New Left movement, in association with the previous category, the anti-nuclear movement was primarily born of the desire to shut down power plants in Western Europe (Whyl and Brokdorf in West Germany and Creys-Malville in France). Also, it attempted to promote the importance of the nuclear energy debate in European politics. It was one of the most radical, unconventional political players in the New Left environmental movement, with an emphasis on unco-ordinated, decentralised actions. On the whole, it did not attempt to play neo-corporatist, formal organisational and partisan politics.

Box 3.2

Case study: the anti-roads movement, UK

The modern, British anti-roads movement began in the 1980s but became most conspicuous in the early 1990s. It was the most prominent part of the environmental movement in all of Europe for the rest of the decade. It was motivated by the aggressive road-building programme put in place by the Tory government during the 1980s and 1990s, which changed the basic structure of the once renowned English countryside with huge, high-speed motorways, and saw an immediate increase in Britons using cars. By 1992 there were 20 million cars in Britain, ten times as many as there were in 1952. Most significantly, a third of this massive increase in numbers occurred during the 1980s.

The concerns held about cars are varied, with some environmentalists believing they are leading us to *carmageddon*. Firstly, the dominance of the car and other petrol/gasoline powered road vehicles has led to profound changes in the very fabric of towns, cities and lives. By accepting that car travel is normal, cities have rapidly expanded into the countryside, and the form of the car-dependent city is a sprawling one. Streets have become less safe for pedestrians, and communities themselves have broken down, producing isolation and loneliness. 'Car culture' has also produced massive environmental problems, with the major impact that automobiles have on climate change due to CO_2 emissions (Newman and Kenworthy 1992). Smog from cars directly threatens human health through contributing significantly to upper-respiratory diseases like asthma, and cars also have direct lethal impact through collisions. Also significant is that high-energy, oil-dependent cities mean countries are heavily reliant on the politics of Middle East oil. Much money and lives have been spent and lost in recent wars waged, in part, to control the flow of these oil-fields.

Road networks were being extended within British cities to provide the fastest travel time for commuters, with the rationale that time savings lead to economic growth. So as well as refuting the powerful cultural icons of cars and oil, anti-road environmentalists were directly rejecting the measurement of all natural goods through the lens of economic rationalism. In this way it rocked the foundations of British society.

Concerned citizens lobbying for the cessation of road building in their area formed a coalition called AlarmUK, and in the years between 1989 to 1993 had limited successes through public inquiries in some anti-roads campaigns such as Preston, Crosby, Hereford, Norwich and Calder Valley. What was unusual was that NIMBY (Not-In-My-Back-Yard) protestors who experienced increasing dissatisfaction with mainstream democratic processes, were joined by more radical 'outsider' activists who had bigger political pictures concerning road-building and who were willing to take direct action against construction sites (the core of these activists being British Earth First! groups). This was a powerful combination of previously disparate voices, and numbers swelled remarkably.

In 1992 the first and definitive campaign of the direct action networks was at Twyford Down in opposition to the proposed M3 in Hampshire. For twenty years there had been extensive public opposition to the building of a three mile section of the M3 which would desecrate two monuments and two more sites of special scientific interest (SSI). After the legitimate channels under British and European law were exhausted and the major players (the Twyford Down Association and Friends of the Earth) abandoned the protest, a grassroots group of protestors established a camp on the second SSSI:

> Calling themselves the Dongas Tribe after the ancient routes that crossed the Down, their number quickly swelled to nearly 100 people . . . The group successfully disrupted construction using various forms of non-violent action . . . By August, more than 30 demonstrations were held involving between 4 and 500 people – an unlikely alliance of Tories, travelers and eco-radicals.
>
> (Rawcliffe 1995: 32).

The group of unlikely bedfellows attracted the interest of the British media and British public. Even after the Dongas were evicted, direct action continued almost daily throughout all of 1993, becoming more piecemeal in 1994 when the road extension was finally built.

Despite ultimate defeat, it became the signature campaign for the next five years, and led to the blossoming of the movement with protests starting in Glasgow, Jesmond Dene in Newcastle, the M11 in east London and Solsbury Hill, and the M65 in Lancashire. From 1995 and 1996 another huge protest was mounted in Newbury in Berkshire. In 1997 action moved to Devon with protests against the A30 (Wall 1999: 65). Later that year, attention was turned to preventing the building of Manchester's second runway (Griggs and Howarth 1999). Although declining somewhat in intensity, the anti-roads movement has continued to be active into the new millennium, and to some extent evolved into the 'Reclaim the Streets' street carnival phenomenon held in cities around the world.

Source: Adapted from Doyle, T. (2005) *Environmental Movements in Majority and Minority Worlds*, New Brunwick, New Jersey and London, Rutgers University Press

There is a major difference between the kind of environmental politics embraced by these two different movements or networks. The nature conservation movement sought 'protection within the existing economic order' (Lowe and Goyder 1983; Rohrschneider 1991: 254). This movement 'proposes reform' by playing politics with a 'green tinge'. The political ecology and anti-nuclear movements demanded systemic change, placing ecological and social objectives above economic concerns. In more recent times, the anti-roads movement in the United Kingdom (see Box 3.2), and European movements against genetically modified organisms (see Box 3.3) have continued and re-invigorated the political ecology traditions.

Box 3.3

Comparative case study: the anti-GMO movements in India and Germany

The birth of genetic engineering in the 1970s has been a controversial subject in most countries, and no aspect more so than the use of biotechnology in agriculture and food production. Genetically Modified crops are now growing across the globe, most of which have been engineered to be pest or herbicide resistant, or a hybrid of both. Around 25 per cent of these crops worldwide are modified with a gene taken from the toxin *Bacillus thuringiensis* (Bt), a naturally occurring soil bacterium that is closely related to the anthrax used in biological weapons. The remaining 75 per cent of GM crops have been engineered to be herbicide-resistant, by the same companies that primarily produce commercial herbicides (Ho, Ching *et al.* 2003: 8–9).

In the minority-world country of Germany, the anti-GMO debate mainly centres on the risks to human health when GMOs are consumed in food products. This follows in the wake of several other food scares such as mad cow disease and salmonella and *E.coli* contamination. Thus a key platform of the movement is to insist that products containing GMOs are adequately labelled. Accountability of corporations and governments and choice for consumers are two key principles championed by the movement, making it a much more consumer-driven opposition to that which exists in India. European activists are also concerned about the harmful effects the *Bt* toxin could have on the biodiversity of the natural ecosystem, in particular the exposure of rare butterflies and moths to GM crops, and effects on soil organisms. Yet in majority-world countries such as India, activists are worried about species loss from GM crops because of the effect it has on food sovereignty and security itself.

The Indian anti-GMO movement emerged from existing activism against global financial institutions, enforced development policies, and the Green Revolution that brought multinational agri-chemical companies to the South (Buttel 2005: 314). The movement can be understood within the broader framework of resistance to globalisation and the neocolonial activities of multinational corporations, with Gandhi's concept of self-reliance constructed to mean independence from the imperial domination of seed companies. Shiva explains how seed represents the entire struggle against GMOs in India:

> The seed has become for us the site and symbol of freedom in the age of manipulation and monopoly . . . In the seed, cultural diversity converges with biological diversity. Ecological issues combine with social justice, peace and democracy.
>
> (quoted in Herring 2005: 205)

Activists in India are worried about food security and the loss of control or sovereignty over their food supply altogether, given the country's struggle against poverty and inequity. Partly because of the GM revolution, agricultural production is now often concentrated on non-food commercial crops grown for export, such as cotton, while people across India go hungry. Furthermore, crops which are not traditional food staples in much of the majority world (such as GM soybeans and corn) are now replacing diverse varieties of legumes, beans, millets, wheat and rice. Finally, farmers and activists are worried that pests will develop a tolerance to the toxin gene, therefore requiring a greater use of pesticides and increasing the cost of production for farmers. Following the introduction of *Bt* cotton, around 500 farmers committed suicide in one year, apparently because of the exorbitant levels of debt they faced due to the additional cost of GM seeds and pesticides (Shiva 2006).

The movement in India is spearheaded by farmers and groups representing the interests of rural people, in particular the Karnataka State Farmers' Association, while in Germany there is a greater proportion of concerned consumers and urban activists. However the tactics both use are not so dissimilar, with the bulk of the movement focusing on holding protests and raising awareness, lobbying governments, and sometimes resorting to radical direct action such as sabotaging laboratories and GM trials.

Source: Adapted from Helman, Z. (2006) 'Opposition to GMOs: A Comparative Case Study of Social Movements in Germany and India', postgraduate paper presented in the School of History and Politics at the University of Adelaide

The nature conservation movement accepted current distribution patterns of both power and economic resources. Both the political ecology and anti-nuclear movements argued for resource conservation along with a more equitable distribution of those resources. The characteristic style of politics of these two groupings was substantially different. Despite the fact that activists within these movements recognise and use, often in mutual criticism, the differences between these political traditions, the public sees both as just part of 'one environmental syndrome' (Rohrschneider 1991: 251–66).

Environmental movements of Eastern Europe: post-industrial movements

The division between Western and Eastern Europe, prior to 'the end of the Cold War', was a key division between liberal democracies with capitalist economies and those countries whose politics were based on authoritarian, state-centred models of socialism.

Before the mid-1980s, there was no environmental movement in Eastern Europe to compare with those in the West. Under authoritarian regimes it is difficult to measure the extent of support for any social movement. There are always movements that remain more or less 'underground' during this period of harsh rule, ready to emerge and flourish when conditions improve. It is interesting to note that it is the non-institutional form of politics that sustains movements under repressive regimes. As discussed in Chapter 1, few NGOs, political parties or other formalised centres of opposition are tolerated in these regimes. Instead, informal networks, groups and associations provide the lifeblood of the social movement's existence. There are no easily defined leaders, no easily located epicentres, and no clear patterns of association to be suppressed. Szabo, in the context of Hungary's greens, writes of this 'structure' of politics:

> The Hungarian ecology movement was born in the mid 1980s, but, given the constraints of mobilizing under the one-party state, it never attained an integrated organization. Rather, it existed in the form of unconnected local citizens' initiatives, single-issue groups, and alternative lifestyle communities. Unlike ecology movements in France and West Germany in the 1970s, there was no unifying antinuclear group in Hungary, despite Chernobyl and the scandals surrounding the only Hungarian atomic power plant in Paks.
>
> (1994: 292)

Major problems in Eastern Europe include air and water pollution, waste management and soil contamination. These are extremely severe in the 'triangle of pollution': the former GDR (Democratic Republic of Germany, also known as East Germany), the former Czechoslovakia and Poland (Schreiber 1995: 365). These countries are directly responsible for a great deal of the pollution in the North, Baltic and Black Seas. Obviously, this level of contamination poses dramatic health risks.

In certain regions of the former Soviet Union such as Lake Baikal, the Aral Sea and industrial centres like Kuzbass, the incidence of cancer 'is up to 50 percent higher than the Soviet average, and respiratory-tract illnesses are common' (ibid.: 364). And the levels of environmental degradation and resulting contamination are simply not known, let alone addressed, in the many Asian regions of the former USSR.

Many of these problems are attributed to the *heavy* industrialisation that took place after the Second World War. While Western Europe was involved in developing advanced 'end-of-pipe' technologies, the governments of Eastern Europe ignored environmental costs. Some writers argue that it was the 'command control' decision-making structures of the 40-year-old Stalinist regimes that were responsible for the lack of environmental regulation and reforms (Fagin 1994: 479–94). Whatever the causal factors, severe environmental problems of this kind leave little space for nature conservation or wilderness concerns, although these also exist in the region.

As the control of the Soviet Union collapsed, environmental movements emerged more openly in Eastern Europe to address pressing problems of environmental degradation and exploitation (however, the experience of Bosnia, a former socialist country, was quite different – see Box 3.4). These movements did not emerge simply as a logical response to these problems, serious though they were. Just as important was the fact that environmental discourse was, in part, tolerated by conservative, communist regimes in the initial period, as the 'political dimension to ecology was unforeseen by the regime' (ibid.: 480). This initial tolerance of environmental arguments allowed other forms of dissent to manifest themselves. Most accounts of dissent in Eastern Europe treat it as a struggle between the values of the one-party regimes and the 'heroic' forces of democratisation. Szabo supports this point but makes another when he contends:

> On the one hand, social forces have 'unlimited possibilities' for articulating new issues because the inertia of official politics recasts any challenge in terms of the broader drama of democracy versus authoritarianism. In Eastern Europe, the result was the existence of some very limited initiatives that had public and intellectual significance disproportionate to the small number of supporters. These small, powerless groups could become capable of articulating very important – even crucial – but neglected sociopolitical issues. On the other hand, mobilization under authoritarian systems is hampered by the administrative-bureaucratic environment and the use of legal and illegal means of social control.
>
> (1994: 287)

Box 3.4

The Bosnia–Herzegovina experience

Bosnia's environmental problems have emerged from three phases of its recent existence: socialism, war, and post-war neo-liberal reconstruction. Socialist economic management and political organisation remain significant in explaining the attitudes of officials towards regulation, accountability, citizen consultation, and the management of natural resources. The 1992–5 civil war has left specific ecological burdens, with the deliberate destruction of natural resources, population displacement and the use of landmines. It also completely destroyed the economy and deterred foreign investment for over a decade, and the country is now reliant on foreign donations for both reconstruction and environmental recovery. However there are fears that the standard neo-liberal economic solution based on attracting foreign investment by opening up the economy, will in due course exacerbate the destruction of natural resources due to an absence of effective regulation.

Bosnia is a unique Eastern European country in that an environmental movement has not emerged through any 'heroic moments' in Bosnian politics. The external funding after the civil war was directed to the creation of civil society groups, but the result has been the creation of the kinds of organisations best suited to meeting the rubric of funding regimes (largely the European Union) and their attached conditionalities. These are groups of technical specialists who tend to avoid political controversy and debate over environmental principles and have stronger ties with funding agencies than with the Bosnian public. Although political authority is fragmented in Bosnia, the policy that drives reconstruction mistakenly assumes that Bosnians never had a capacity to mobilise on issues such as the environment, despite the earlier history of environmentalism under the Yugoslav regime. Grassroots environmental campaigns can work in Bosnia, but these are not the groups that get funding.

Source: Adapted from Fagan, A. (2006) 'Neither 'North' nor 'South': The Environment and Civil Society in Post-conflict Bosnia–Herzegovina', *Environmental Politics*, 15, 5: 787–802

In the early phases, aided by the new tolerance of the 1980s, environmentalists were well received by the broader *populus* as champions of democracy and dissent. In the case of Hungary, the protest against the building of an Austrian-funded dam on a Hungarian section of the Danube provided a symbolic epicentre not only for the 'ecologists', but for many other forces of 'democracy' (ibid.). Environmentalism has been used for many disparate purposes. On this occasion, it was used to undermine an authoritarian communist regime. Interestingly, as the 'velvet' revolutions took place, green forces became accepted as 'legitimate' and were quickly absorbed into the mainstream politics of the new quasi-pluralist systems. Many green interests were accommodated in the new political parties (see Chapter 5). With the demise of state socialism, most people in the East saw more pressing issues than those they narrowly construed as environmental. They were eager to move to the 'free-market' systems of the West, and growth and democratisation were portrayed as the key to the East's success. The arguments of post-industrial theorists ceased to be heard. The environmental movement had served its purpose in the transition period. Chatterjee and Finger observe:

> Not surprisingly, in a highly politicized society and at a highly political moment, the Green movement in the East was first and foremost a political movement with political, i.e. national agendas. Be it in Hungary, Poland, Czechoslovakia, or Estonia, the Green movements turned rapidly into green parties, which in turn quickly acquired a share of national power. But once this had taken place, the Green movement declined. In retrospect, it turns out that the Green movement in the East was instrumental in the transition of the Eastern European countries to a market economy. Yet despite enormous ecological problems facing the East, in the 1990s the Green movement has substantially lost momentum.
>
> (1994: 66–7)

In a paper by Szabo (with Kereyi 2006), they analyse environmental mobilisation in Hungary over a far larger period of time, from the period of initial mobilisation until the present. At first glance, it seems that Chaterjee and Finger's conclusion is apt, but with a longer timeframe now available for analysis, the story is more complex, and less pessimistic about the longevity of green politics under post-communism. After the birth of the movement had taken place under the previously analysed period of 'illegality', the remarkably high 'movement capital' evaporated soon after the regime change. The new post-socialist national framework provided limited resources which resulted in the emergence of narrowly focused NGOs, while in the final most recent phase the intensification of transnational interaction gave rise to new mobilisation patterns and increased action by the environmental networks (Kereyi and Szabo 2006).

The fight for Zengö Hill (see Box 2.2) is evidence that vibrant green movements are still capable of emerging when political opportunity structures are in place. In 2004 new groups and more radical ideas emerged through a protest campaign to protect a valued hill (Zengö) from being damaged for a NATO military site. This campaign included urban environmental activists with ties to global justice groups in other countries and the Hungarian branch of Friends of the Earth, which was able to use its international network contacts to put pressure on NATO and the EU. It could be said that the parts of the Hungarian environmental movement that participated in the Zengö campaign increased their social movement dimensions (Doherty and Doyle 2006). Some networks within the green movement have not been co-opted by mainstream politics and have remained a vibrant though informal voice in Eastern Europe. This level of non-institutional politics may not be as 'visible' or as 'reportable' as its more institutional counterpart. It is fluid, amorphous and flexible. Those concerned by the evidence of environmental degradation continue their efforts to change the way in which these new regimes respond to this difficult part of their legacy. In addition, transnational environmental activism has provided new networks for support outside the nation-state.

Finally, in many ways Szabo's initial story remains salient: the idea that green politics could be construed as post-industrialist, central to questions of political economy; so central, in fact, to the business of deciding the actual make-up of political regimes that it seemed to contradict those then numerous post-materialists who argued that ecopolitics was the thing wealthy people – usually from the West – dabbled in after their societies had achieved their basic elements/needs for Maslowvian survival.

Divisions between movements of North and South: post-colonial movements

One of the trends to emerge from green globalisation has been for Northern groups to attempt to incorporate those very different concerns expressed by their Southern compatriots. There is no doubt that increasingly, diverse interpretations of green identity are under growing pressures to homogenise. But we must accept that, despite these recent trends at conceptualising and sharing grand green narratives, the actual environmental issues on the ground are profoundly different in the South from in the North. Movements, therefore, which surface in countries like India, Bangladesh, Chile or Somalia – in the majority world – will be more oriented around issues of environmental security: that is, the rights of people to gain access to the fundamental resources for survival: air, water, earth (food sovereignty, biodiversity) and fire (energy). For example, the most pressing environmental issues in the Indian Ocean region, a region comprising one-third of all the peoples on the Earth, are almost all anthropocentric: people fighting for food sovereignty and water security; struggling for adequate admission to a market which provides healthcare and adequate shelter; providing a society to live in which is not consistently ravaged by wars – such as Afghanistan, Palestine or Angola – and wave after wave of colonialism, in all its forms, creates catastrophic tsunamis of human making – (Doyle and Risely 2008). Whereas, in many parts of the North these issues, however compassionately understood, are literally worlds apart from the lives of the wealthy minority, most of whom will only ever experience the lives of the majority – what Toffler (1970) calls the Living Dead – through the vicarious experiences offered by travel and lifestyle programmes on television and the internet.

Unlike some forms of post-modern and post-positivist analysis, then, we still find the binary mega-division between majority and minority worlds – though imperfect – a useful one, as it continues to match and describe the 'empirical reality' as we have encountered it; as long as it is understood that these great divisions are neither necessarily geographically oriented, nor nation-state specific. Rather, there is an immense gulf in the context of comparative environmental movements between the experiences of the majority of the earth's people (the South), when contrasted with those encountered by a small minority (the North). A rather simple, often quoted, equation needs to be spelt out here. Approximately 80 per cent of the Earth's resources are either consumed

Plate 4 ***Indian protestors at the 2004 World Social Forum, Mumbai. Courtesy of Joel Catchlove***

or owned by approximately 15 per cent of the Earth's people. On the other hand 85 per cent of the Earth's people have access to only 20 per cent of the Earth's resources.

Although wilderness-oriented movements do exist in the South, they are largely overshadowed by debates about the environment/development nexus (Khan 1994).[5] For example, in South Africa there has been a long history of campaigns to preserve wilderness areas and to construct the large 'game parks' and natural reserves such as Kruger National Park and Umfolozi (Carruthers 1995; Player 1997). But wildlife conservation, the creation of game parks, national parks and nature reserves can have a significant impact on the livelihood of local populations, many of whom do not welcome these developments. Given a view of wildlife conservation, of wilderness and a tension between the needs of humans and wild animals, the building of parks and reserves has frequently been based on the removal of people from the areas intended for conservation (Player 1997). Although there are examples where this has not happened, removal is frequently the basis for wildlife conservation and the wilderness experience. This can be seen quite clearly in the case of South Africa, accentuated by both colonial and apartheid assumptions about race and land. The struggle to create Kruger as a national park is instructive (Carruthers 1995). Initially the park was formed to provide against game depletion for future hunting, and local populations were incorporated into the expanding park with their grazing, gathering and hunting activities restricted. At a later stage these populations were removed and the success of conservation was directly based on the hardship of dispossession. Subsequent expansion of Kruger was based on the (often multiple) dispossession of local populations. As apartheid was replaced, the question of land, dispossession and restitution moved onto the political agenda. Many white conservationists feared that the new government would abandon conservation in favour of desperately needed economic development to support improved standards of living for the newly enfranchised majority. This has not been the case. The first real test was protecting St Lucia, then under threat from proposals for sand mining, when the new government reaffirmed its commitment to nature conservation for the benefit of all South Africans (McEachern 1997). Nonetheless, the urge to remove people to make national parks took a while to lessen. For instance, the creation of the Richtersveld National Park on the arid coast of Namaqualand was initially predicated on the dispossession of a group of people who had managed to carve out a meagre existence after relocation as part of the normal politics of apartheid. After a vigorous local struggle, a new agreement was struck so that the local community could remain, with restricted rights but with some access to the income stream from tourist visits to the area.

Since the end of apartheid there have been efforts to build better relations with the displaced communities living near the parks. Restitution of lost lands with leaseback and managerial rights, rather on the Australian model for Uluru, have become more common. Nevertheless, the normal pattern remains and little has been done to lessen the tension between conservation and local well-being in the areas surrounding the park activities.

Obviously, the post-materialist thesis is largely meaningless in the context of environmental activism in the South. There is little about 'higher values' (in Western terms) when considering the South's environmental crisis, and crisis it is. Also, little credence is given to post-industrialist arguments, as most people in the South see the key problem as lack of ownership of their own resources. Chatterjee and Finger comment on this point:

> [For] the Third World Network in Malaysia or the Centre for Science and the Environment in India, it is no longer industrial development *per se* which is considered destructive of the environment. Rather it is the fact that development remains controlled by the North instead of the South. The weakness of this argument, of course, stems from the fact that it mixes together Southern peoples and Southern Elites.
> (1994: 77)

Also, many parts of the South are not heavily industrialised economies, although this is changing rapidly.

The dominant view of the North is that the poverty of the South has caused and continues to create environmental degradation.[6] This environmental degradation is of grave concern to the North, now advocating global ecology and, as a consequence, seeing itself as inevitably having to share the Earth's essential survival mechanisms with the South. Along these lines, the North portrays the major problems of the South as deforestation, species extinction, global climate change, desertification and overpopulation.[7]

The unwritten assumption here is that the South is the main environmental offender, while the North is a model of environmental controls and reforms. In this view, the North sees itself as bringing its environmental message (including that of sustainable development and good management) to the South, to save the latter from itself. With increased growth and democratisation, 'civil society' will emerge, the North argues, promoting conditions where people will help themselves.[8] It is true that Northern environmental activists, mostly through the vehicles of international NGOs (discussed in the next chapter), are active in the South. Nonetheless, many successful environmental networks in the South are inspired by local activists, many of whom would be construed as 'poor' by Northern standards. They are not degraders or sustainers, but actors.

In the Philippines, for example, a vast and vibrant local environmental movement is fighting environmental degradation head-on across a range of fronts (Broad, 1994: 813). In Africa, grassroots organisations have taken a leading role in the struggle against environmental degradation (Ekins 1992: 114). In India, the movement against the Narmada Dam (ibid.: 88; Roy 1999) (which threatened to dispossess entire communities of their traditional lands) and the Chipko movement were mostly driven by local activists. The case of the Chipko movement illustrates how a group of local people, with only the power of their own solidarity, were able to curtail logging in Uttar Pradesh in April 1973 by hugging trees, using the Gandhian method of *satyagraha* (literally 'firmness in the truth', but usually translated in the West as non-violent or passive resistance). This spontaneous movement spread to many parts of the Himalayas over the next five years (Ekins 1992: 143). In South America, there are few better examples than that of Chico Mendez and his battle to fight for the lifestyle of his community of rubber-tappers against development interests. In Mexico, the Zapatista uprising, championing the indigenous Chiapas, formed in direct opposition to corporate and government adoption of the North American Free Trade Agreement in the mid-1990s (see Box 3.5). In Thailand, local people fought the Nam Choan Dam project in the 1980s. Few Western and Japanese investments had been so strongly resisted by the local population since the 1950s. The construction of dams in Thailand had previously led to deforestation, changes in local climatic conditions, declining soil fertility, and degraded water and fishery resources (Hirsch and Lohmann 1989: 445). In the first half dozen years of the new millennium, Thai environmentalists have shifted their energies towards the construction of Thai–Burma, transcontinental gas pipelines which, again, are actively displacing the powerless from their homes (Simpson 2007). Likewise in China, during the first emergence of environmentalism in the 1990s, activists have

Box 3.5

Zapatismo: a revolutionary green social movement

The Zapatista uprising against the Mexican government occurred on 1 January 1994, the day the North American Free Trade Agreement came into operation. This was at a time when the government and a few corporations extracted much wealth from Chiapas, and yet indigenous groups experienced some of the worst poverty extremes in Mexico. For example, roughly 30,000 people in Chiapas died the year before the uprising of hunger and diseases relating to malnutrition (Chiapaslink 2000: 13). While the government proclaimed that NAFTA would attract foreign investors and elevate Mexico to the status of a First World country, the Zapatistas believed it would be a 'death sentence' for their communities which were mostly made up of traditional, poor maize farmers (Kingsnorth 2003a: 19). They took their name from the Mexican revolutionary hero Zapata's fight for the common ownership of land and the right of small farmers to control their own villages in the Mexican revolution, as Zapata's *ejido* system of land tenure that guaranteed land for peasant and indigenous farming populations was reversed as part of the NAFTA negotiations.

The uprising consisted of 12 days of fighting and the seizure of San Cristobal de las Casas and 5 Chiapas towns before a unilateral ceasefire was declared. To date there have been no satisfactory peace talks as the government has been waging a low-intensity military conflict ever since which has not aided dialogue, and the Zapatistas are not interested in reintegration into the Mexican state as a political organisation. The movement is not an attempt to seize state power, the idea is to dissolve power to the level of communities to give the indigenous people control over their economic and political lives, and enable them to determine how their resources are used. In direct opposition to the colonialism of globalisation, the Zapatistas have seized back land for self-sufficiency in subsistence farming, and trade through direct bartering and local collectives (Esteva 2003).

The rebellion has sought to organise the community as a whole, across sectors and generations, as a social movement. Their masked spokesperson Subcomandante Marcos and others are convinced that these 'free spaces, born of reclaimed land, communal agriculture, resistance to privatization, will eventually create counter-powers to the state simply by existing as alternatives' (Klein 2002: 220). To spread the message, it has adopted the internet and independent global communications systems as a key medium to alert the world to its presence and spread the revolution:

> We have a choice. We can have a cynical attitude in the face of media, and say that nothing can be done about the dollar power that creates itself in images, words, digital communication, and computer systems that invade not just with an invasion of power, but with a way of seeing

> the world, how they think the world should look. We can say, well, 'that's the way it is' and do nothing. Or we can simply assume incredulity: we can say that any communication by the media monopolies is a total lie. We can ignore them and go about our lives. But there is a third option that is not conformity, nor skepticism, nor distrust: that is to construct a different way, to show the world what is really happening, to have a critical world-view and to become interested in the truth of what happens to the people who inhabit every corner of the world.
>
> (Subcomandante Marcos 1997)

The message at the heart of Zapatismo is a global call to revolution that tells supporters to begin wherever they stand with their own 'weapon', whether it be a video camera, words, or ideas; to create spaces in which other spaces are possible, to create 'a world in which many worlds are possible'.

The Zapatista revolution intersects with the global justice movement in more than just the struggle against neoliberalism and economic colonisation, their very form mimics the internet itself. Power in the network is decentralised, allowing for 'immense freedom, creativity, and innovation on the part of each local centre and each individual contributor, while each still remains connected to a network for the sharing of ideas, visions, analysis, research, stories, and even hardware' (Nogueira 2002: 294). It is not simply a localised indigenous uprising; rather, it is intended as a revolution of ideas with international impact.

fought against the colossal Three Gorges Dam project on the Yangtze River, despite government silencing and repression (Coonan 2006). In fact, the list is endless. Poor people are most affected by environmental degradation in the South, and they are the key actors and inspirations of the environmental movement.

Broad writes:

> A reader need only flip through any issue of the British journal *The Ecologist* or the Malaysian *Third World Resurgence* journal to find numerous case-studies of the poor being involved in protecting the environment – replanting trees, struggling against the enclosure of ancestral lands, fighting for indigenous and community resource management.
>
> (1994: 813)

There are also ongoing struggles for food sovereignty and security in the face of market-based global food production systems, and also to retain culturally appropriate and valued foods. The threat of introduced Genetically Modified Organisms (GMOs) highlighted by Southern movements are also faced in the North, yet Northern greens take a decidedly different approach (see Box 3.3).

It can be shown time and time again that jeopardising the right of the poor to subsist often leads to environmental activism in difficult circumstances. Democratisation can create a better climate for environmental action but, as the case of the Ogoni in Nigeria shows, even repressive military regimes – and now democratic regimes in transition from military control – face challenges from dedicated and brave poor people seeking to defend their environments.

Although responses to some forms of continued poverty do have a negative impact on local environs,[9] both the degradation and the poverty are generally caused by ancient land use and ecological histories (such as deforestation, desertification), which are rendered damaging when coupled with the more recent (in human terms) exploitation of the South by the North, and the complex interplay between these factors.

So on the political level, these problems are often the result of many years of imperialism on behalf of the North. The nations of the South were seen as a treasure trove for Northern traders. Now that these nations are 'independent', they have embarked on their own development projects. Over the past generation, there have been three distinct periods of development in the environmental movement (see Box 3.6).

This pursuit of development has been accompanied by environmental degradation for a number of reasons. Damage is partly the result of local industries necessarily producing 'dirty' products in a bid to maintain competitiveness in the 'new global economy', recently deregulated under the General Agreement on Tariffs and Trade (GATT). Multinational and transnational companies, however, are the principal environmental offenders in emerging economies. They can produce commodities far more 'efficiently', using cheap labour and with the less stringent environmental demands of local legislation. These companies also continue to transport their industrial and toxic wastes to these 'developing' nations. Consequently, in 'newly industrialising economies', such as Hong Kong, between 75 and 90 per cent of its factories are illegally dumping liquid wastes into the territory's waters (Douglas *et al.* 1994: vii). Between 1.5 and 5.8 million of Hong Kong's residents have experienced health problems due to air pollution. In Bombay, and other parts of India, the air quality produced by industrial wastes is among the worst in the world. Its antiquated water supply system and lack of

Box 3.6

Three stages of development in Southern environmental movements

Stage 1: 1960s

The first development decade in the South brought optimism. Northern-style growth and development were the goals. So much so, that there was little movement opposition within the countries of the South. Movements from outside the South – mainly in the form of large NGOs – occasionally entered the political sphere.

Stage 2: 1970s

During this period, environmental movements emerged in the South. Again, these were dominated by some key NGOs, e.g. the Green Belt Movement in Kenya, the Environment Liaison Centre International, Environment and Development Action in the Third World, and Sahabat Alam Malaysia. Few movement participants opposed the Northern development ideology, but they fought for 'people's development' (another development), not governments' or multinationals' development. This type of development shares many similarities with the political ecology movements in the North, particularly Western Europe.

Stage 3: 1980s to the present day

During this period, movements split into two categories. After a period of emphasising local and grassroots development in the 1970s, many networks in these movements began to collaborate with government and international agencies again, as in the 1960s. Many coalitions of grassroots groups and local NGOs formed umbrella coalitions e.g. Asia Pacific People's Environmental Network, African NGOs Environmental Network, the Asian NGO Coalition, etc. Many of these powerful coalitions, bypass government on occasions, and negotiate directly with international aid agencies. The other category was the development of environmental protest movements, very similar to the political ecology movements of the North. These networks criticised Northern development schemes. They criticised Northern science and technology, the industrial practices of transnational corporations, national governments, Northern governments, and international aid agencies.

Source: Adapted from Chatterjee and Finger (1994)

a proper sewerage system mean that diseases such as typhoid, malaria, asthma and even bubonic plague are rife in the rapidly expanding suburbs, again reaffirming the nexus between environmental degradation and poverty.

Transnational environmental movements: globalisation and the green public space

Transnational conflicts over power and ideology have been central from the beginning of the new environmentalism of the 1970s, in particular since the debates about the risk and limits of growth. But while global in its analysis, there was very little evidence of global environmental protest action or of movements working effectively across borders. Prior to the late 1990s, most national environmental movements lacked either the time or money to be able to engage in consistent international activity. However, transnational environmental action is on the increase, with one study demonstrating that there are now 150 more transnational social

Plate 5 ***The Narmada dam, Gujarat, India. The Narmada dam has been the focus of a struggle lasting some two decades, between the pro-development forces championed by the Indian Government and the World Bank, and the new environmental NGO, the Narmada Bachao Andolan (NBA) composed of environmental and human rights activists, scientists, academics, writers and those directly affected. NBA is directly committed to stopping the numerous dam projects in the Narmada Valley. Author's private collection, April 2000***

movement organisations (TSMOs) in the environmental field than there were in 1972 (Bandy and Smith). Environmental groups now make up 17 per cent of all TSMOs (2005: 16).

Although environmental and social movements often break through geopolitical boundaries, they are still premised on nation-state assumptions. Yet within this 'bounded innovation' of national experience, there is an increasing body of evidence that campaign repertoires for resistance have transnational origins. The tree-sitting escapades in Grenoble in 2003–4 echo similar tactics used by the anti-road protestors in the UK in the 1990s, which in turn echo similar tactics used in North America and Australia in the 1980s. In turn, these passive resistance tactics echoed the Gandhian tree-huggers of Chipko in the 1960s, later manifesting themselves in Burma with Buddhist monks ordaining trees, and then in Hungary, when Catholic priests blessed national flags atop Zengö Hill (Doherty and Doyle 2006).

The political forms of social and other mass movements are suited to crossing boundaries. Often boundless, anarchistic, ambiguous, many-faceted, they lack the defined edges of more institutionalised political bodies. For example, where governments of nation-states appear increasingly unable to monitor and trace the activities of transnational corporations, social movements, due to their jelly-like structurelessness, seem more capable of 'border osmosis', of oozing through largely non-institutionalised pores in the fabric of frontiers, tracking and contesting the rapidly expanding power of transnational capital (Doyle 2005).

The social movement of transnational environmentalism has intersected with the 'global justice movement', a term denoting the broad movement which emerged after several turn-of-the-century coordinated global actions against the meetings of transnational institutions such as the World Trade Organisation, the International Monetary Fund and the World Bank (most memorably, the 'Battle of Seattle' at the 1999 WTO meeting). Uniting human rights activists, environmentalists, unionists, farmers, workers, queer rights activists, feminists and many others, from the outset it has demanded that the South gain fair and equitable access to resources; it has challenged the over-consumption patterns of the North, whilst also critiquing ways in which Northern development interests act to maintain the dependency of the South. It is deeply opposed to the global neo-liberalist agenda which, it argues, is destroying people and the environment in pursuit of ever greater corporate profits.

These new transnational global justice networks of protest movements and social forums have been inspired by the evidence about unaccountable and unjust transnational sites of power. They are also facilitated by several new developments: first, the break-up of the USSR has removed the tendency to see all post-1945 conflicts in bi-polar terms; second, cheaper communications in the form of phone and fax, the internet and email (see Box 3.6), have transformed the ease of transnational exchange of information and co-ordination of action. Also vital has been the reduced cost of air travel in facilitating face-to-face networking and developing relationships of trust and solidarity, as embarrassing as it is for environmentalists, given the contribution of air travel to CO_2 emissions in the current climate debate.

The main actors within the global justice movement are globally mobile activists possessing the 'cultural capital' of higher education (often inclusive of English language skills) and 'the social capital inherent in their transnational connections and access to resources and knowledge' (20). This "movement of movements" (Torgerson 2006) has the power to unite voices from the formerly colonised regions of the planet together with the former colonisers, and there is evidence that Southern movements are increasingly driving the global green movement agenda (Doyle and Doherty 2006). Elements of these transnational networks, however, have also proven themselves capable of reproducing the very inequalities they are fighting against, for example in constructing grand omnipresent narratives – such as climate change – which gut the local stories and realities of the global South.

Conclusion

There is no single entity that is the environmental movement. There are only many and varied environmental movements, with many and varied networks within and between them. Some of these green movements can be referred to as social movements; some cannot. Also, new social movement (NSM) models may not always be valid in the context of explaining the post-colonial movements of the global South. In each country, in certain regions, in some cultures, these mass movements – a green public sphere – reflect the dominant cultural and economic aspirations of their national societies, while others project themselves in opposition to these dominant values. In addition to national and regional distinctions between environmental movements, it is important to understand that since the late 1980s there have also existed transnational environmental movements that, like the transnational corporations that they either fight or support, no longer have a 'fixed address'. To complete this account of the character of diverse environmental movements, it is necessary to turn to the most formal and visible components, the non-governmental organisations.

Further reading

Chatterjee, P. and Finger, M. (1994) *The Earth Brokers: Power, Politics and World Development*, Routledge, London and New York.

Doherty, B. and Doyle, T. (2007) *Beyond Borders: Transnational Environmental Politics*, Routledge, London.

Doyle, T.J. (2000) *Green Power: The Environment Movement in Australia*, University of New South Wales Press, Sydney.

Doyle and Risely (2008) *Crucible for Survival: Environmental Justice and Security in the Indian Ocean Region*, Rutgers University Press, NJ, NY, London.

Ekins, P. (1992) *A New World Order: Grassroots Movements for Global Change*, Routledge, London.

Kalland, A. and Persoon, G. (eds) (1998) *Environment Movements in Asia*, Curzon, Surrey.

4 Green non-governmental organisations

- **Constitutionalised relationships**
- **Pluralist, corporatist, authoritarian and post-modernist theories of NGOs**
- **Characteristics and structure of green NGOs**
- **Transnational NGOs**

Introduction

Non-governmental organisations (NGOs) are the most visible players in environmental politics around the globe. They are involved in many different spheres of politics, from the local community level, through the politics of the nation-state, to international politics. They exist in both the predominantly non-institutional domain of social movement politics and in the institutionalised milieu of political parties, administrative systems, governments and beyond.

Although the term non-governmental could include the commercial 'private' sector, the label 'NGO' is rarely applied to business (Bebbington and Thiele *et al.* 1993: 5). Nonetheless, business may well sponsor sets of NGOs that may make green claims while defending business interests in the policy process.[1]

During the 1970s and 1980s, there was an explosion in NGO numbers. Porter and Welsh Brown estimate that by the early 1980s there were approximately 13,000 environmental NGOs in 'developed countries', 30 per cent of which had been formed in the previous decade. Approximately 2,230 NGOs operated in the 'developing countries', 60 per cent of which had emerged in the same decade (Porter and Welsh Brown 1991: 56).

Even though these figures are impressive, growth in the NGO sector jumped even more startlingly in the mid to late 1980s. For example, the

African NGOs Environment Network (ANEN) formed in 1982. Twenty-one NGOs were among the founding organisations, but by 1990, membership had swelled to 530 NGOs in 45 countries. In Latin America and the Caribbean, there are currently over 6,000 NGOs, most of which have formed since the 1970s. India has more than 12,000 'development' NGOs, Bangladesh has 10,000 environmental NGOs, and the Philippines has some 18,000 (Princen and Finger 1994: 1–2).[2] As was discussed in Chapter 1, even in one-party states there has been an eruption of environmental NGOs beginning in the 1990s and increasing into the first decade of the new millennium. In China, for example, total numbers of green NGOs have grown from zero to over ten thousand, largely due to concerns of citizens in relation to their country's unprecedented, rapid industrialisation (Economy 2004).

This rapid increase in NGO numbers reflects an explosion in environmental activism. Most NGOs have constitutions, and these symbolic documents give the organisations more permanence than many grassroots groups and informal networks. Hence NGOs are fairly stable entities and their actions can be more closely tracked and counted throughout a given period. For every NGO there are many more informal groups, associations, coalitions and networks. For example, for the 12,000 Indian NGOs mentioned above, there are 'hundreds of thousands of local groups' operating in India (Princen and Finger 1994: 2).

Also interesting was the growth in membership numbers within certain individual Northern NGOs during the 1980s. At this time the mass media was beginning to pay increased attention to international environmental issues, such as the Bhopal and Chernobyl disasters, the discovery of the hole in the ozone layer, and developing scientific consensus over climate change, which contributed to widened public interest (Keck and Sikkink 1998: 128). From 1985 to 1990, the membership of Greenpeace (an Amsterdam-based international NGO) increased from 1.4 million to 6.75 million. From 1981 to 1992, Friends of the Earth (FoE) (an international NGO originating in the USA in 1969) more than doubled its number of member groups. The Sierra Club (a national US organisation) increased its membership from 346,000 to over half a million between 1983 and 1990. Often these dramatic increases in membership are associated with similar increases in expenditure (Princen and Finger 1994: 2–3).

There is some evidence to suggest that there has been a decline in memberships of large green NGOs operating in the North during the

1990s (Doyle 2000). For example, Greenpeace was substantially criticised for its 'unorthodox' direct marketing techniques and consequently its membership has fallen. A number of other environmental organisations ran into policy difficulties and were exposed to fairly sustained media criticism, often supported by business. The sharp recession of the early 1990s, with increasing unemployment, also put pressure on environmental organisations as economic issues received greater attention. It is also possible that the 1990s decline in northern NGO membership tallies may just be a levelling off after two decades of expansion. In the first decade of the twenty-first century, NGOs memberships is on the rise again in the minority world with the climate change issue attracting large numbers. In the South, numbers continue to escalate dramatically as governments struggle more and more to address urgent issues of environmental remediation.

What are environmental NGOs?

The first thing which must be understood is that environmental NGOs are political organisations. An organisation, like a social movement, has distinctive collective properties. The most readily recognisable feature of these organisations is the constitution. Briefly, the constitution is symbolic of both the acceptance of legitimacy and rationality; it publicly states the organisation's intention to work within the dominant structures of the state. Murray Edelman writes:

> On the one hand the constitution legitimizes in morally unquestionable postulates the predatory use of such bargaining weapons as groups possess: the due process of law, freedom of expression, freedom of contract, and so on. On the other hand, it fixes as socially unquestionable fact the primacy of law and of a social order run in accordance with a code that perpetuates popular government and the current consensus on values: the rule of law, the power to regulate commerce, the police power, and so on.
>
> (1964: 18–19)

These constitutions establish certain rules which dictate how power will be dispersed throughout an organisation, how decisions will be made, and how people will relate to each other (see Box 4.1).

Although environmental NGOs are diverse, they are far less heterogeneous than the political groupings spread throughout

Box 4.1

Characteristics of an organisational constitution

1 It is a very broad statement of intentions.
2 It uses symbolic language to attract a greater number of people, and to diffuse opposition.
3 It can be used to either blunt or focus criticism.
4 It acts as a mechanism of power. It can be used as a tool of repression or as protection against those who attempt to coerce others in the organisation.

environment movements, as there are certain shared rules for playing politics. Sometimes, of course, these rules are open to interpretation. The political characters of NGOs are not really indicative of the most radical elements within environmental movements. Many radical activists of a more revolutionary bent remain deliberately removed from this more formal realm of politics for two reasons. First, they believe their more radical goals will be co-opted by mainstream politics. Second, they feel better equipped to fulfil their goals within more fluid, less structured political forms.

As with social movement theories, there are numerous explanations of when, why and how organisations form. Many of these theories about NGOs are oriented around different understandings, or models, of society – with particular emphasis on the role and make-up of the state in a given society. Let us now look at four basic models of society, describing alternate ways in which power is dispersed within them. Each model depicts NGOs in a different light, as they respond to varied forms of state power: pluralism; corporatism; authoritarianism; post-modernism.

Interest groups and lobbying: a pluralist view

The first theory, interest group or pressure group theory, is strongly linked to the pluralist interpretation of liberal democracy. In this view, NGOs are seen as interest or pressure groups. This is rather confusing as we have used the term group, in the previous chapter, to connote a more informal human collective form. But in traditional political science the terms group and organisation have been used interchangeably.

In a pluralist world, it is assumed that there are no significant concentrations of power: power is diffused in society and all citizens have some power resources that can be used to achieve their aims or interests. People will have different aims and ambitions, and different interests and grievances. People with shared visions, grievances or interests are free to form organisations (treated as either 'interest' or 'pressure' groups) to press their claims in the political process. With varying degrees of skill and varying amounts of power, these groups battle for influence. The political system responds to at least some of these demands, and there is a degree of incremental adjustment.

In this model the state is treated as 'neutral', having no particular interests or goals of its own. In pluralist systems NGOs exist to lobby. They do not seek public office directly, but through influence provided by members, public opinion, money and power, they seek change indirectly. Duverger argues that these organisations

> do not participate directly in the acquisition of power or in its exercise; they act to influence power while remaining apart from it; they exert pressure on it . . . Pressure groups seek to influence [those] who wield power, not to place their own [people] in power, at least not officially.
>
> (1972: 101)

Pluralist accounts capture some aspects of what happens with NGO political action. Many environmental groups do form in response to a shared sense of concern about the harmful environmental consequences of a proposed or existing economic development. Once formed, such groups will use the full array of measures to pressure the political systems to either stop or regulate the damaging economic activity. The problem with the pluralist account lies in its assumption about what happens in the competition with other groups and forces as they struggle to influence government. Not all groups are equal either in their access to power resources or in their access to the political process. Some environmental NGOs do have large sums of money to spend on their campaigns, do employ scientific and legal experts and do have great skill in putting their case using the existing hostile, sceptical media.

Greenpeace would be the most obvious and important example, but not many environmental groups are like this. In their contests, these groups are often small, poorer and weaker than their business opponents. In these circumstances, the pluralist framework can produce a misleading interpretation of events. It should also be noted that there are serious problems with pluralist assumptions about the character and role of the state. In environmental disputes, it is very unusual for the state to be a neutral party, seeking to arbitrate between proponents and opponents of a particular project. Quite frequently, government either initiates, fends for or supports a given development, which is then subject to environmental dispute. There is one further point to note about this treatment of NGOs as if they are just pressure groups: to consider environmental NGOs in this limited way takes them out of their movement context and, at best, gives a one-sided account of the dynamics of their actions.

Closer relations with the state: corporatist environmental policy making

Pluralist models of power and politics describe a world in which what the state does is a product of the pressures brought to bear on it by various competing groups and social forces. These groups put pressure on the state but form no part of the state system. In the 1970s and 1980s, such accounts were challenged by the rise of various kinds of corporatist analysis. Initially, corporatist theory simplified the assumptions of the pluralist world. Instead of a multiplicity of competing groups, emphasis was placed on just a few, basically those of business and labour. Attention was then paid to the interactions between business, labour and the state in various kinds of corporatist situation. From this flourished a whole literature, extending the corporatist/incorporatist model to a whole range of different policy issues, including environmental organisations and environmental issues.

In general terms, there are two contrasting evaluations of this change. Some corporatist theorists see this situation as most positive for 'civil society' as the process 'entails the state's recognition of the functional interest groups and their incorporation into the process of policy making and implementation' (Capling and Galligan 1991: 57). Others are concerned that NGOs are threatened by the process and interpret it as an effort by the state to incorporate and neutralise opposition.

In a study of Australian environmental policy over a decade, McEachern identified three different ingredients (see Box 4.2) that make up this corporatist process: incorporation, assimilation and adaptation.

McEachern's ingredients, although derived from the Australian context, would seem to apply to the ways in which environmental policies

Box 4.2

Three ingredients in corporatist environmental policy making

Incorporation

Environmental activists and the business community are brought together inside a set of 'normal' political negotiations defined by the nation-state or a combination of states. These negotiations establish the sense that there is a consensus position on environmental concern. In the act of negotiation, dissident sections of the environment movement and business (that is, those who do not believe consensus is possible and/or desirable) are constructed as just that, dissidents against an emerging, shared, politically acceptable position.

Assimilation

Second, there is the process in which the socially critical discourses of ecology and environmental concern are taken and turned into legitimate, acceptable, non-threatening discussions about existing economic and resource development practices. The initial formulation of 'sustainable development' was of this kind.

Adaptation

This ingredient necessarily complemented assimilation in that it involved a consideration of the evidence of environmental damage drawn from the arguments of environmental concern. This evidence of damage could then be assessed by relevant governments and scientists and incorporated into policy considerations. Although it would have been possible for politicians to marginalise environmental concern, a potentially more effective response was to accept the evidence of damage and to seek policies that would allow maximum economic growth while minimising damage.

Source: Adapted from McEachern (1993: 180–1)

have been shaped during the 1980s and 1990s in a variety of liberal democracies. Even at the international level, in those forums sponsored by the United Nations, this description of the corporatist policy process seems quite apt in explaining the relationship between green

NGOs and nation-states and their shared efforts in environmental policy making.

For example, Chatterjee and Finger, in *The Earth Brokers*, describe a comparable policy-making scenario at the United Nations Conference on Environment and Development held in Rio de Janeiro in 1992. Many NGOs were invited to the conference. Accordingly, many observers equated this with more power being given to the NGOs by the participating nation-states. Some were admitted into the process of formulating the policy documents, while others participated in an alternative NGO gathering in the same city. Despite the environmental NGOs being allowed to take part in the debates over the wording of *Agenda 21*, Chatterjee and Finger believe that the process of corporatism weakened their political capacities. They were now seen as endorsing the policies of UNCED. In turn, this could be used to legitimate the activities of big business and Northern governments. Opposition and dissent were marginalised, extracted and discarded from the policy debate, by selling the notion that the economic imperative of growth and development (sustainable development) would actually enhance the environment. Chatterjee and Finger observed:

> This, in our view, was the result of a long-term transformation of the Green movement world-wide, combined with the very way the UNCED process was set up, as a means of reducing potential protest by feeding people into the Green machine . . . if there was no substantive outcome in terms of conventions and documents, UNCED was at least an exercise in mobilisation and cooptation, weakening the Green movement on the one hand while identifying and promoting potential opponents – mainly from the South – on the other.
>
> (1994: 100, 103)

The more general debate over corporatism and incorporation suggests the limits of these arrangements for environmental NGOs. On the one hand, moving into such corporatist forums gives an NGO access to information, funds and significant decision makers. Here is a chance to be influential and have some input into the content of important environmental decisions. On the other hand, there is the danger of incorporation. The symbol of an environmental NGO's co-operation may be of far greater value to government than any specific decision made. In some cases, the ability of an NGO to respond to environmental problems may be impaired by taking part in a corporatist project. Such participation has costs in terms of time, resources and energy that cannot then be spent on direct mobilisation.

NGOs and authoritarian regimes

So far we have only looked at the relationship between NGOs and the state within liberal democracies, yet the activities of NGOs are certainly not limited to these. Environmentalism is present in non-democratic regimes, but the regime type dictates to what extent these NGOs can challenge state practice and which green ideas filter through. Radical green ideas in authoritarian states are accessible mainly through satellite broadcasts and websites run by dissident exiles.

In China, after the student uprising and events of Tiananmen Square in 1989, the Chinese Government was concerned that NGOs provided a political sphere where intellectuals, workers and peasants could meet to stir up additional unrest. From 1995–97, the government placed a two year suspension on the registration of new NGOs, and, in September 1998, the State Council established a set of restrictive 'Regulations for Registration and Management of Social Organisations' (Economy 2005). In effect, all organisations had to have government approval, with no right of appeal if unsuccessful. In the new millennium there have been significant gains made in the NGO, or third sector, with environmental groups being at the vanguard of this change. In fact, there are now over 280,000 NGOs in China, with green NGOs amongst the most visible (Knup 1997: 10). Some of these green NGOs include: Friends of Nature, Green Earth Volunteers, Green Watershed, Green River, Nu River Project and the China Rivers Network. This movement, although not challenging the power of the state directly, is at the forefront of political and democratic reform. Many members of these NGOs have leaders well trained in the law and journalism, which provides them with access to important institutions to advance their cases. The influential Beijing journal *Economics* has called the anti-dam movement in China the 'new social power in China' (Fu Tao 2005).

As with the case of China, in recent times, the Islamic government in Iran has moved from a position of banning NGOs to a point where they are now encouraging environmental NGOs as a means of improving environmental governance. Still, in an official sense, they are not permitted to work as political groups in any way that might challenge the regime. When Iran held a major international environmental conference in 2005, it did not invite any NGOs from Iran or abroad in case they asked questions that the regime did not want to answer. These environmental organisations function as a release valve for parts of civil society, and are seen as a convenient way to 'downsize government' and reduce its direct responsibilities in environmental policy-making (see Box 4.3). However, the process of providing access to the national political system enables Iranians to communicate their dissatisfaction with the regime, and while far from revolutionary, it is potentially emancipatory. It has seen the emergence of a green public sphere in Iran, which could in time be a harbinger of increasing democracy (as was the case in Hungary – see Chapter 3), ultimately leading to the overthrow of the ruling theocratic regime.

The military dictatorship of Burma, on the other hand, has less effectively controlled state territory and public life, and has no participation in legalised environmental mobilisation like that of China or Iran. Open dissent is not tolerated by the ruling SPDC, and only a very limited number of NGOs are permitted, those which engage in non-threatening,

Box 4.3

Green politics in Iran

Six reasons the regime tolerates green NGOs

1. Environmental problems are serious and NGOs can act as useful partners, often with technical expertise.
2. Due to public sector downsizing, NGOs are seen as economically efficient service providers.
3. Iran has a very young demographic profile and some avenues of participation must be conceded. It is then best to embrace moderate environmental NGOs and try and control them.
4. Many young women run or operate out of these NGOs and it is seen as a non-threatening avenue for female public participation.
5. Green concerns are viewed narrowly, in post-materialist fashion, relating to issues such as pollution and biodiversity. Debate over environmental justice and issues of political change are not permitted.
6. The emergence of green NGOs may not denote the emergence of democratic reform but, rather, the construction of a state-controlled civil society.

Source: Doyle, T. and Simpson, A. (2006) 'Traversing more than Speed Bumps: Green Politics under Authoritarian Regimes in Burma and Iran', *Environmental Politics*, 15, 4: 15

community-development activities. The regime has exploited its gas reserves and directed environmentally and socially catastrophic energy export projects, the best known of these being the construction of the Yadana gas pipeline to Thailand in the 1990s which involved forced labour, forced relocation, rape, village and crop burning, and summary execution of the ethnic Karen people (Karen Human Rights Group 1996). Both the environmental and human rights abuses have been strongly and militantly resisted by these ethnic and religious minorities, and green politics has merged with struggles for survival within a repressive state. This nexus between environmental protection and human rights has led to the development of a new discourse of 'earth rights' (Doyle and Simpson 2006: 6). The only options for dissent in such a regime are military insurgency or the voicing of grievances through international NGOs.

Post-modern/post-structural theories of NGOs

The first three theories revolve around the central importance of the relationships between NGOs and the state. The fourth, a post-modern theory of NGOs, sees the state as just one actor – however important – in a complex interplay between disparate actors. In this model of society, the state is no longer seen as the major pivot or central conduit of mediation between actors: rather, actors deal more directly against or in concert with other actors. In fact, in these more post-structural models, the state is increasingly being bypassed by NGOs (see Figure 4.1). For example, rather than simply hoping to influence the state to change legislation governing the private sector, environmental groups and organisations now adopt a variety of approaches: some directly lobby corporations associated with environmentally irresponsible practices; some work within them in an attempt to reform these firms from 'within'; or alternatively, others work directly against them in attempts to diminish consumer or investor confidence in that company.

In general terms, post-modernists see the world as fragmented and in a situation of dislocation and crisis. Just as the post-materialist theory of environmental movements comfortably sits alongside pluralist perceptions of politics, the theories of post-industrialism are more easily accommodated into a vision of a post-modern polity.

There is now an intense fragmentation of political processes and actors, as evidenced by the description of the palimpsest presented in the previous chapter. Let us briefly look at the post-modernist theory of NGOs posited

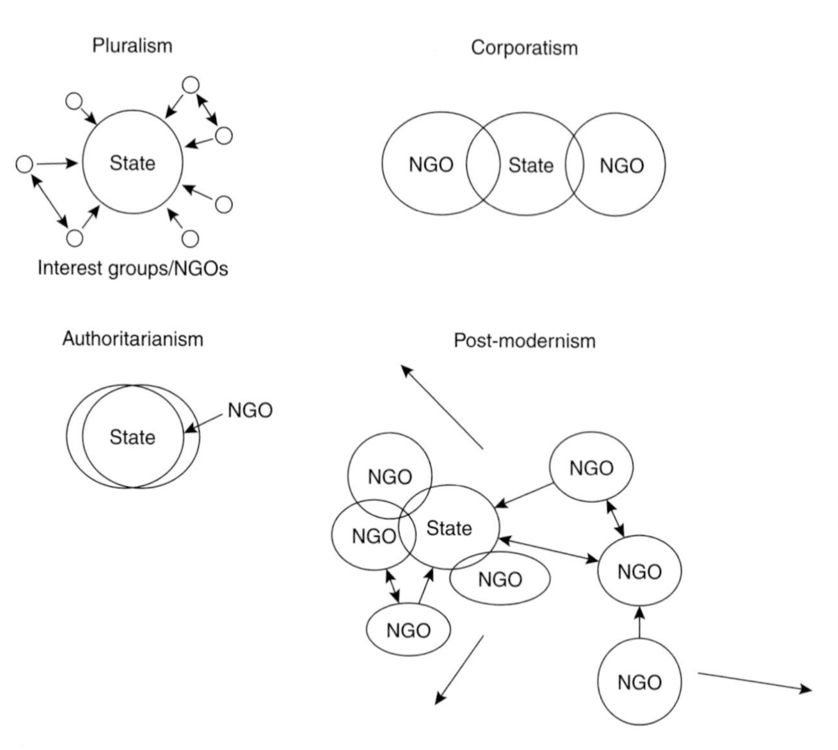

Figure 4.1 ***Pluralist, corporatist, authoritarian and post-modern theory: relationships with the state***

by Thomas Princen and Matthias Finger. Unlike the pluralists and the corporatists, Princen and Finger believe that there is an environmental crisis, which demands a new form of politics as a response. As well as all the standard economic and technological changes that transform politics, Princen and Finger note:

> In addition, the global ecological crisis has reinforced and accelerated this process toward post-modernism, rather than reversed it or slowed it down. It has led to more fragmentation, further eroded collective projects, and contributed to the multiplication of social environmental actors.
>
> (1994: 61)

They argue that the project of modernity is under attack and there is an 'absence of a common reference point'. Nation-states, particularly within any globalised economy and facing global ecological problems, have lost the capacity to be effective in solving environmental problems. More than that, nation-states are often dangerous to the Earth's ecology. Their primary functions are twofold: first, to provide a basis for an expanding military–industrial complex; second, to promote the notion of a national economy, based on 'growth' and 'progress' at all costs. Princen and Finger argue that the search for effective and substantial environmental reform has to be pursued in a domain beyond and below the nation-state level. Their predominant answer is to value the style of NGO politics that has emerged from this fragmented and diffused political situation.

> As environmental NGOs free themselves from traditional politics, change the reference point and privileged means of action, grow in numbers and inter-connectedness, and become increasingly transnational, they contribute to societal change and transformation in another way: they become agents of social learning . . . Indeed, rather than focusing on traditional politics, how to influence it and how to mobilize for it, environmental NGOs build communities, set examples, and increasingly substitute traditional political action. They become agents of social learning, whereas social movements were actors of political change only.
>
> (ibid.: 65)

While much of what they write evokes the non-institutional experience of social movement politics, Princen and Finger see NGOs as largely separate from social movements. They contend that it is not only the traditional defenders of the modernist project but also the social and mass movement theorists who remain bewitched by the nation-state model of politics. They are right about this point, and many NGO theorists remain equally transfixed by pluralist and corporatist theories of access to the politics of the nation-state. The model of social movements discussed in Chapter 3 is not confined to what goes on inside nation-states but does include NGOs in the environmental movement. The real advantage of Princen's and Finger's account of separate and distinct environmental NGOs is found in its contrast with pluralist assessments. Whereas the pluralist, corporatist and authoritarian models emphasise lobbying for incremental change, post-modernist theories, like Princen's and Finger's, see NGOs as possessing far more direct, creative and transformative powers. Further, post-modernists see social movements and environmental NGOs forming as a result of fundamental ruptures within the modernist project, whereas the pluralist 'interest group' theorists, for example, regard them as a minor variation on the politics-as-usual theme.

Depending on which of these four models of power is used to explain the emergence and definition of NGOs, there are different answers to key questions about green NGO politics. Each model presupposes very different notions about what NGOs actually do and what tasks they usually perform, and each promotes a clearly separate view of the relationships between NGOs and the state. Finally, each framework gives a different perspective on the internal dynamics of NGOs. It is now useful to discuss several broad criteria that can provide a preliminary typology of green NGOs practising all around the globe.

Typology of green NGOs

Although not as diverse as environmental movements (as they share common organisational characteristics), environmental NGOs vary on the basis of an array of determining factors. These include their geopolitical origins; their political ideology; their size; the level of their political focus; their funding sources; what they provide (what tasks they actually perform); their internal politics/structure; and their relationships to the state. Let us briefly review the first five characteristics with reference to Tables 4.1 and 4.2.

The geopolitical origin denotes the region in which the organisation either formed or operates. The 'place of origin' is important, as many large organisations, particularly those involved in global ecological issues, increasingly see themselves as transnational. Some organisations, although originating in the North, are major operators in the South. In more recent times, Southern NGOs have also established extensive global networks.

With reference to Table 4.1, there is a basic difference between the political ideology of radical and reformist organisations (as discussed in Chapter 2). Also, under these broad headings there is a myriad of eco-philosophies: from political ecology to resource conservation; from deep ecology and eco-feminism to social ecology; from the left to the right. Obviously, in different parts of the world certain NGO ideologies will be more apparent.

Green NGOs vary dramatically in size. For example, Green Don (Zedon) (Table 4.2), a Russian NGO operating in the Black Sea, has 40 members and no operating budget (Global Env. Facility Black Sea Env. Programme 1995), whereas Greenpeace International (Table 4.1), with nearly 3 million members, had a 2004 income of 38.9 million euros (Doherty 2006: 6).

Some organisations are involved in policy making at a global level; some aim at national politics; some are regional and some are intensely local,

Table 4.1 ***Typology of green NGOs in the North***

NGO characteristics	*Greenpeace*	*Friends of the Earth*	*Australian Conservation Foundation*	*Japan Tropical Forest Action Network (Jatan)*	*Earth First!*
Geopolitics	North, North in South	North, North in South	North	North, North in South	North
Internal structure/ politics	Hierarchic, top-down	Dispersed confederation of regions	Federation of state chapters and branches	Centralised	De-centralised
Political ideology	Radical political ecology	Radical political/ social ecology	Reformist, resource conservation	Reformist, resource-conservation	Radical deep ecology
Size	2.8 million financial members	71 member groups, 1.5 million members and supporters	15,000 members	4 staff, vast network	Cannot measure
Political level	Transnational	Transnational	National	International/National	Regional
What they do	Non-violent direct action, mass mobilisation	Mass mobilisation	Lobby appeal to elites	Mass mobilisation	Sometimes violent direct action
Relationship to nation-state	Few	Few	Close	Few	None
Funding source	Membership drives, direct marketing	Membership, private	Membership, government, direct marketing	Network funding	Newsletter, low costs

Table 4.2 ***Typology of green NGOs in Eastern Europe and the South***

NGO Characteristics	*EFT*	*Green Don*	*Grameen*	*BRAC*	*DAWN*
Geopolitics	East West in East	East	South	South, North in South	South
Political level	National	Local, regional	Regional, national	National, international	Regional, national
What they do	Conferences, scientific research	Public education	Providing env./dev. funding	Env. and dev., widespread direct provision	Women's community env. issues
Funding source	70% int. World Bank	None	Investors	International, government	International community

forming to resolve, for example, an environmental issue indigenous to a tiny community catchment area.

Many NGOs generate their own funding through membership dues. Others are more aggressive in the marketplace, soliciting money directly from the public in the form of donations, bequests and merchandise sales. These direct marketing techniques are more prominent in the North. Some NGOs, in both the North and the South, also receive money from their national governments. In the South, many NGOs receive money from Northern NGOs and donor agencies, often bypassing their national governments. This sometimes creates tensions when certain NGOs are seen by governments as 'workers for foreign interests'.

For the remainder of the chapter, let us consider in more detail the final three differential variables: what do green NGOs actually do; what are their internal politics; and what are the impacts of globalisation on green NGOs.

What green NGOs do: North and South

Neo-pluralist models are far more appropriate in describing the role of NGOs in the North than in the South. NGOs lobby in numerous forums. Sometimes they lobby local or national politicians directly, by using their potential electoral powers. On other occasions, NGOs lobby administrators in government departments, who are often more 'permanent' than their elected counterparts. At times, particularly in the world of international affairs, lobbyists work at influencing diplomats or scientists. Others try to influence the policies of political parties and corporations or try to convince a certain community that one environmental action is better than another. Finally, many NGOs lobby other NGOs, trying to get some agreement on green objectives and strategies.

Brian Martin refers to this emphasis on lobbying as 'the appeal-to-elites' method (Martin 1984: 110–18). In appealing to elites, environmentalists have to speak a similar political language to those already inhabiting the halls of power. Table 4.1 refers to the Australian Conservation Foundation (ACF), probably the most important mainstream green NGO in Australia. During the years of Labor Party rule (1983–1996), the ACF worked very closely with government and it was widely seen as politically influential and politically aligned in a party sense. NGOs working in a close relationship with government necessarily define their environmental goals in terms that make sense to the *status quo* and, in turn, the *status quo* operators can gain entry into the environmental movement.

Although pressure group – or grassroots style – politics is the dominant form of politics played by green NGOs in the North, it would be wrong to assume that it is the only form of politics. In true corporatist style, many NGOs are increasingly delivering local environmental projects, funded and often set up by a combination of governments, corporations, financial institutions and donor consortiums. In Australia, for example, governments and business interests now fund 'community organisations' which form to service government grant schemes (largely funded from the ongoing sale of Telstra). Such groups as Landcare, Coastcare, Tidy Towns, Bushcare, Our Patch and Local *Agenda 21* associations are cheap, on-the-ground service providers for these government schemes and programmes. These groups are closely controlled and harnessed by bureaucratic and industry initiatives and arrangements. Providing the nomenclature of 'community' to these service providers deliberately disempowers grassroots community activists. On many occasions, these groups do valuable work. But they are *not* community groups and organisations as have been defined in this work. They are not part of the tradition of the modern environment movements which genuinely emerged from community politics. In some ways they could be more aptly referred to as 'communeaucracies', or 'bureaumunities' (Doyle

2000). In the previous chapter, and in this chapter's final section, we prefer to refer to them as green governance groups (GGGs).

In the European political ecology movement, there are many actions that are aimed at mass mobilisation rather than appealing to elites. Sometimes mass mobilisation is used as an advanced form of lobbying. On other occasions, mass mobilisation strategies are based on the assumption that more widespread change is needed than is assumed under the pluralist model. These changes include alterations in mass consciousness and individual value systems. As a result, some of the more radical NGOs in the North involve themselves in mass education programmes.

The direct actions of Greenpeace (Table 4.1), augmented by shrewd utilisation of mass media, were designed to attract widespread support from the global citizenry, rather than gaining favour directly with elites. Earth First!, a US direct action NGO (which refers to itself as a 'non-organisation') steeped in the ideology of deep ecology, is similarly engaged in this form of protest (Table 4.1) (Foreman and Haywood 1997).[3] Unlike its Northern counterparts, however, it does not rule out militant action (most specifically property damage) in its efforts to 'protect Mother Earth'. This method of eco-sabotage is almost uniquely an American phenomenon. Most direct-action NGOs remain committed to the principles of active non-violent resistance (Doyle 1994a).[4]

Obviously, this more radical, mass-mobilisation role of NGOs is more readily explained by the post-modernist model of politics. In this time of fragmentation and dislocation, some NGOs are seen as deliberately bypassing governments and acting more directly at local, regional and global levels. But acceptance of this post-modernist model cannot be automatically equated with more radical forms of politics. Sometimes, by rejecting government links, NGOs are also entering the more socially conservative world of the marketplace. The appeal-to-elites method of lobbying governments is replaced by appeals to corporate elites. On some occasions, there may be little to separate the demands of business and the goals of the state, but it is still useful to distinguish between them. For example, instead of lobbying governments with a hope that they will monitor and control more closely the deeds of large corporations (through legislation), some NGOs in the North are dealing directly with these corporations. Chatterjee and Finger discuss this increasingly close relationship between big business and some of the less radical Northern NGOs:

> WWF, for example, received $50,000 each from oil companies Chevron and Exxon in 1991. The National Wildlife Federation conducts enviro-seminars for corporate executives from such chemical giants as Du Pont and Monsanto for a US$10,000 membership fee in their Corporate Conservation Council Programme. The Audubon Society meanwhile sold Mobil Oil the rights to drill for oil under its Baker bird sanctuary in Michigan, garnering US$400,000 a year from this venture.
>
> (1994: 70)

The funding arrangements of the largest environmental 'governance' organisations are increasingly intermeshed with the vested interests of transnational capital, leading to strong market competition between them (see Box 4.4). These more free-market interpretations of environmental reform are culminating in some NGOs no longer seeing themselves as non-profit organisations but as 'players' who can trade freely in the marketplace. This has led, in some circles, to growing concern about the democratic accountability of NGOs (see Box 4.5).

Box 4.4

Global Green Governance NGOs: World Wildlife Fund, Conservation International and The Nature Conservancy

In a controversial article published in World Watch by MacChapin, the funding arrangements and subsequent conflicts of interest of the (US branches of) the three biggest international conservation organisations are examined. It damningly concludes that as corporate and government money flow into these organisations, their programmes are 'increasingly excluding, from full involvement in their programmes, the indigenous and traditional peoples living in territories the conservationists were trying to protect' (17).

While overall there has been a sharp decline in money available for conservation programmes since 1990, the funding available to these 'big three' has increased due to the expansion of their fundraising reach into new areas, such as private foundations, bilateral and multilateral agencies (such as the World Bank), transnational corporations, the US government (and other national governments), and individuals. Corporations they tap into include Chevron Texaco, ExxonMobil, Shell International, Weyerhauser, Monsanto, Dow Chemical and Duke Energy. The combined revenues of these three organisations increased from US$635 million in 1998 to US$899 million in 1999 to $965 million in 2000. The largest of these, The Nature

Conservancy, now has assets in excess of US$3 billion. Their total investments in conservation in the developing world have grown from roughly US$240 million in 1998 to nearly US$490 million in 2002.

In their mandate to preserve nature areas (often referred to as the 'fortress conservation approach'), it is argued that these organisations preference objective ('apolitical') Western science over indigenous knowledge, and shut indigenous people out of areas where they have always lived in order to preserve non-human nature. Indigenous peoples and conservationists frequently have very different agendas, which clash excessively when these organisations and their corporate investments are involved:

> Conservationist agendas . . . often begin with the need to establish protected areas that are off-limits to people, and to develop management plans. If they include indigenous people in their plans, they tend to see those people more as a possible means to an end rather than as ends in themselves. They are seldom willing to support legal battles over land tenure and the strengthening of indigenous organisations; they consider these actions 'too political' and outside their conservationist mandate. They have been reluctant to support indigenous peoples in their struggles against oil, mining, and logging companies that are destroying vast swaths of rainforest throughout the world.

The article suggests there has been a shift within the 'big three' away from building local capacity (by helping launch local NGOs to work with the indigenous communities in their own countries), towards seeing themselves as semi-permanent organisations that don't want to work themselves out of a job.

All three of these organisations responded to Chapin's article with heated criticisms of inaccuracies and false assumptions. Unacceptable factual errors aside, they argued that the whole premise of the article was incorrect and that they have depended on partnerships with local indigenous peoples to conserve critically threatened ecosystems. However, while Conservation International claimed to work with industry to 'change private sector operational practices at a global level while having positive impacts for people and biodiversity at the site level', for example working with Starbucks through the Conservation Coffee™ programme to pay higher wages to environmentally friendly coffee farmers, at no point do they point to examples where they have listened to indigenous communities and campaigned against the destructive practices of the corporations they are sponsored by. At no point do they claim to have disrupted 'business-as-usual' and caused any shift in consumer or corporate thinking which is arguably essential for global conservation.

Source: Adapted from Chapin, M. (2004) 'A Challenge to Conservationists', *World Watch*, Worldwatch Institute, November–December, pp. 17–31

Box 4.5

Accountability within NGOs

There is a growing body of literature which examines the accountability of NGOs, given that most NGOs demand the transparency and democratic accountability of governments and corporations, and that some NGOs are becoming increasingly larger and more powerful.

According to NGO Watch (www.ngowatch.org), which was set up by the conservative think-tank American Enterprise Institute and the Federalist Society to mirror the Corporate Watch NGO (www.corporatewatch.org),

> this growing local and global role [of NGOs] has in large part been unchecked and unregulated. Coupled with sparse (or reluctant) practices of public disclosure and a spate of high-profile NGO scandals in the last decade, calls for greater transparency in NGO operations have been resounding. Who funds NGOs? How effective are their programs? How do they influence governments and international organizations? What are their agendas? And to whom are they accountable?

While NGO Watch was set up to explicitly undermine the influence of NGOs through exposing their 'scandals', there is some merit in demonstrating that not all NGOs are 'forces of good' in the world.

Brown and Fox (1998: 439) define accountability as holding individuals and organisations responsible for their performance, something usually enforced by vertical hierarchical power relations. However, with hierarchical authority largely absent within environmental coalitions, alternative modes of monitoring and assessment need to be forged by NGOs and the actors within them. While this is challenging within networks which hold an unequal distribution of power and resources by members, power relations need to be balanced and mutual influence processes in place so that all members can be accountable (440). Brown and Fox believe that as a minimum, coalition members should be able to

> define their expectations for performance, monitor compliance with those expectations, and successfully press for changes in behaviour that does not meet minimal standards.
>
> (442)

Ebrahim (2003) has examined the current accountability practices of development NGOs using three criteria, upward–downward, internal–external, and functional–strategic; observing that accountability in practice has emphasised 'upward', 'external', and 'short-term functional' at the expense of the others. While environmental NGOs have vastly different outcomes, they are similar in that there is not enough emphasis on long-term strategic processes, or downward accountability to the more marginalised members who are supposedly being 'empowered' in environmental justice coalitions.

With increased income derived from deals, investments and sales of organisational products, some Northern NGOs are privately buying tracts of land for wilderness preservation and other nature conservation purposes.[5] Also, Northern NGOs are directly funding governments and NGOs in other countries, most particularly those in the South (but also in the East). One example is the way in which US NGOs are buying the debts of poor nations in direct exchange for protected wilderness areas. Schreiber explains this process:

> US environmental groups have bought approximately $1 million of Bolivian debt for $100,000. This has now been offered to Bolivia in exchange for an enlarged nature protection area and a bill for stricter environmental protection in such areas. Similar (debt-for-nature) agreements have been made with Costa Rica, Ecuador and the Philippines. The first debt-for-nature swap with a formerly socialist country, Poland, is now underway. The World Wildlife Fund (WWF) plans to buy US$1 million of Polish debt for $100,000.
>
> (1995: 378)

Despite the more recent emergence of forms of environmental politics that could be explained by a post-modernist view, NGOs in the North remain primarily a political lobby, still identified as pressure and interest groups, and dominated by a pluralist 'way of seeing'.

In other parts of the world, green NGOs are involved in very different forms of eco-activity. Of course, there are still the lobbyists and the mass educators, but in many nations of the South, NGOs are the direct providers of infrastructure. For example, in many cities in Asia (like Bandung in Indonesia, Mumbai in India, or Bangkok in Thailand) where there is a massive population explosion, entire cities are forming outside of the established 'city limits' (Douglas *et al.* 1994: ix–x). In these cases, NGOs are directly involved in the provision of clean water; the physical labour of cleaning up refuse and the disposal of solid wastes; the building of shelter and the provision of sewerage systems; treating people directly for disease and malnutrition; direct provision of food and other basic essentials for living; and co-ordinating many other 'hands-on' tasks and activities.

Plate 6 ***Slum housing, south of Delhi, India. Entire cities are forming outside of formal infrastructures in the South. Here slum housing boils over, along railway lines south of Delhi, March 2000. Author's private collection***

Some Southern NGOs act as conduits for the purposes of larger Northern NGOs. It is interesting that Northern NGOs increasingly accept the importance of 'community-driven projects' in other countries but do not feel this is important in their own countries. The funding received by Southern NGOs directly from the North has been quite large in the past, although it has been decreasing in more recent times as the South's 'independence' leads it to find new ways of funding its own 'development' projects, coupled with the 'turning off' of funds from the North.

With poverty seen as both a cause and an outcome of environmental degradation, many direct programmes are initiated by Southern NGOs to create an awareness of social advancement, economic emancipation and income generation for the poor.[6] One way that enables poor people to resolve some of their community problems relates to the availability of credit. One NGO operating in Bangladesh, the Grameen Bank (Table 4.2), is an excellent and successful example of extending banking facilities to the poor, and especially women. The Grameen Bank was started in 1976 by Muhammed Yunus (winner of the 2006 Nobel Peace Prize), and has since extended nearly US$1.8 billion in tiny loans for self-employment purposes to over 2.4 million women in over 39,000 village centres (Yunus

1999: 90). The success of Grameen is recognised in the fact that 48 per cent of those who have borrowed for 10 years have crossed the poverty line to the positive side of the ledger, and another 27 per cent have come close (Matthews 1994: 184). Both the UN and the World Bank are now interested in micro-credit schemes, and USAID and the World Bank have pledged over $2 billion to the Grameen Trust which is building replicas in many other countries (ibid.: 185).

Just as the key environmental issues are perceived differently in the South, so too are the roles green NGOs play. In many places, governments simply have not been able to respond to the environmental problems in their countries. NGOs, both local and transnational, have filled this void. Consequently, the neo-pluralist and corporatist issues of 'access to the state' through lobbying are largely redundant in the discussion of environmental policy. Instead, structuralist and, to a lesser extent, anarchist and post-modernist models provide sets of analytical tools that appear far more appropriate in explaining what NGOs actually do in the South: they often de-emphasise the role of the state; they directly target local communities; and they build transnational networks with Northern and other Southern NGOs. Princen and Finger write of the less institutional focus of NGOs operating in the South:

> Just as Northern NGOs are becoming more institutionalized, Southern NGOs are building organizational skills and financial independence and, as a result, increasingly demand greater autonomy and less dependence on Northern supporters . . . As Southern NGOs are becoming more independent and setting the international agenda, Northern NGOs are looking to the South for ideas, as well as to establish their own international credibility.
>
> (1994: 8)

Internal politics and structures in Northern NGOs

There has been much study by political scientists of the relationship between NGOs and the state, but there have been few studies of the internal political dynamics of NGOs. Often internal politics determines the external politics of an organisation: whether it pursues a close relationship with the state and/or certain businesses, or whether it maintains its independence.

To the pluralists, interest groups form to create incremental changes within a political framework dominated by the state. Most are seen as single-issue organisations, not pushing an agenda of systemic political change, as such. As a result, the internal structures of NGOs seeking to lobby governments will often be very similar to governments themselves. Internal structures, both formal and informal, reveal how decisions are made within organisations and how power is distributed. For example, NGOs in the pluralist picture will usually have 'top–down' decision-making structures, with clearly defined leaders and formalised procedures; their administration will be similarly centralised and bureaucratised (with clear differentiation between tasks), and there will be a clear separation between members, elected honorary officials and professional bureaucrats. On occasions, NGOs are so closely intertwined with the state that they base their offices in the same cities where governments centralise their administrations and may mimic national electoral procedures in their own organisations.

Those NGOs whose activities could be explained using post-modernist models of power would remain distant from the state for two reasons. First, they would see the level of nation-state politics as increasingly meaningless in terms of resolving complex global ecological problems. Second, they are crisis-oriented (and often revolutionary by nature) and would contend that playing politics in a mainstream fashion could only lead to incremental changes, which would not be sufficient to solve complex environmental problems.

Obviously, if one were to construct environmental politics in post-modernist terms, one would demand far more from the internal workings of green NGOs. As Princen and Finger would argue, green NGOs have to provide an example to the rest of the world, and this includes how they empower their own people. So rather than reflecting the mainstream political structures (and forging a closer relationship with the nation-state), their internal dynamics must simultaneously challenge *status quo* politics and provide new political pathways.

Internal politics are most visible when particular cases are explored. Two examples of the pluralist pressure-group-style NGOs are the Sierra Club in the United States and the Australian Conservation Foundation in Australia (Figure 4.2). Both organisations work closely with governments and, at times, have been heavily funded by them. Both see themselves as fundamentally lobby groups and educational organisations and their structures reflect this mainstream politics. Both organisations are even structured along federal lines, reflecting the state electoral divisions found within the USA and Australia. Both organisations, although seemingly directed by officials elected by the membership, are dominated by career administrative professionals.

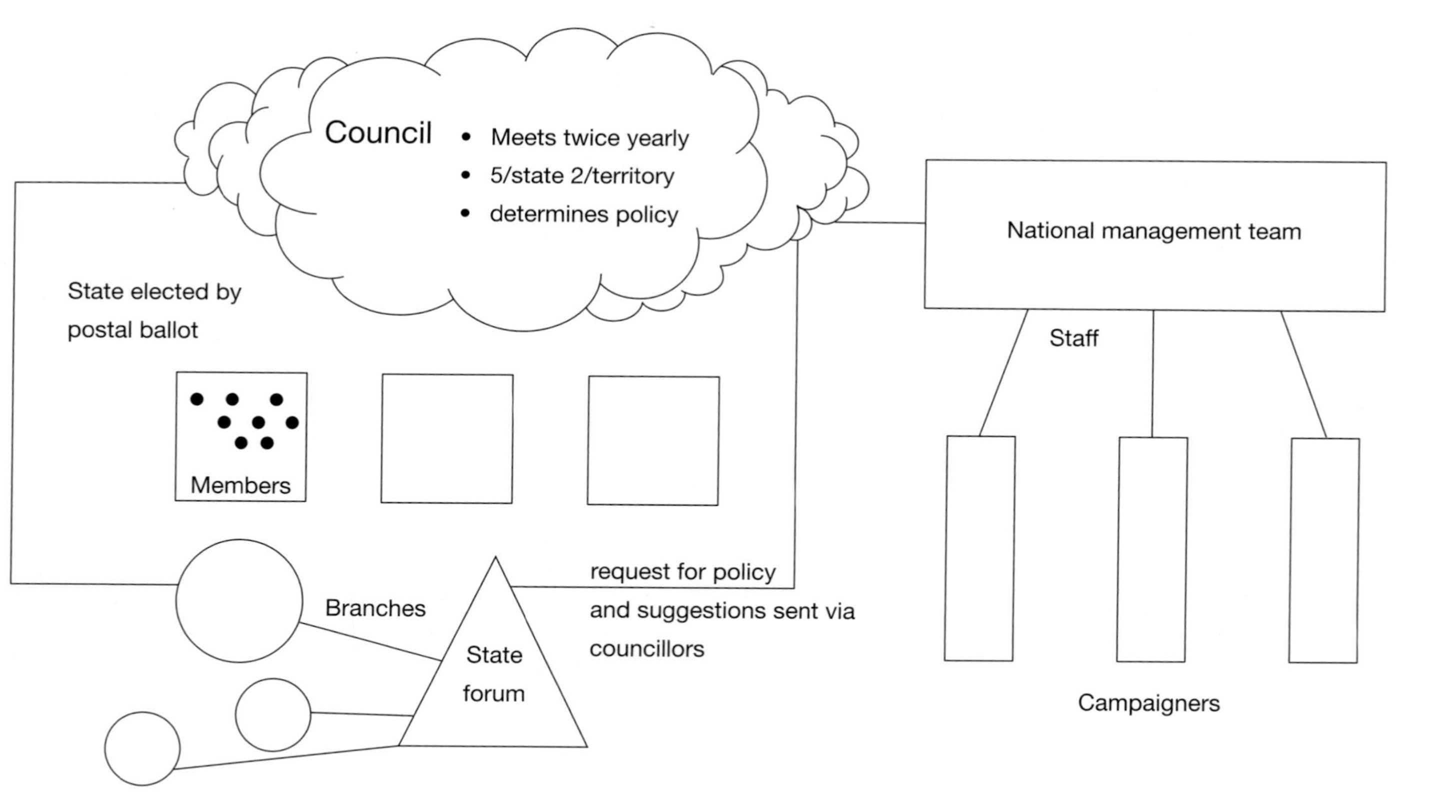

Figure 4.2 ***The Australian Conservation Foundation (structure)***

Friends of the Earth's (FoE's) decision-making structure (Figure 4.3) reflects its political ecology background.[7] FoE seeks to challenge advanced industrial societal ways of living and, consequently, has a more post-modernist perspective on its role as an NGO. As part of this, it promotes participatory democracy. Indeed, participatory democracy is a central plank in many green NGOs. FoE does not possess a huge number of members but seeks to involve its members, as much as possible, in the everyday running of the organisation.[8] Its administration attempts merely to provide co-ordination rather than a centralised determination of policy. The loose confederation of regional groups is given a large amount of autonomy to set their own political agendas. Although FoE is an organisation, it is almost as non-institutional (at least in theory) as an informal group or network. This informality and lack of centralised control reflects a radicalism in its green goals. There is a strong sentiment in these more radical NGOs that if environmental issues are to be genuinely resolved, then these political changes begin at the intra-organisational level.

It would be simplistic to suggest that all NGOs not pursuing environmental changes in partnership with the state are more radical in their internal political form. Greenpeace, for example, is also an organisation which believes that environmental problems cannot be resolved by a direct appeal to government elites. Unlike the political and social ecological organisations, however, it has not traditionally advocated systemic changes. Rather, as evidenced by its direct-action campaigns, it relies heavily on the existing media to broadcast its often issue-specific, mass-mobilisation campaigns. But by viewing its structure (Figure 4.4) one can immediately see that the organisation is highly centralised, with very little room for dissent among its operatives. It models itself less on government (or open democratic structures, like FoE) and looks remarkably like a large corporation. Its members are basically magazine subscribers with almost non-existent powers for influencing organisational policies. Greenpeace responds to these charges from other environmentalists with the hard to dispute fact that 'it gets the job done'.

Whatever the differences in these pluralist, corporatist and post-modernist models, most NGOs experience tensions in occupying two worlds: the non-institutional realm of social movements and the more institutional world of governments. In organisations like the Sierra Club and the ACF, which hold intimate connections with, respectively, the United States Democratic Party and the Australian Labor Party, there is an ongoing debate about being too close to political parties, governments and

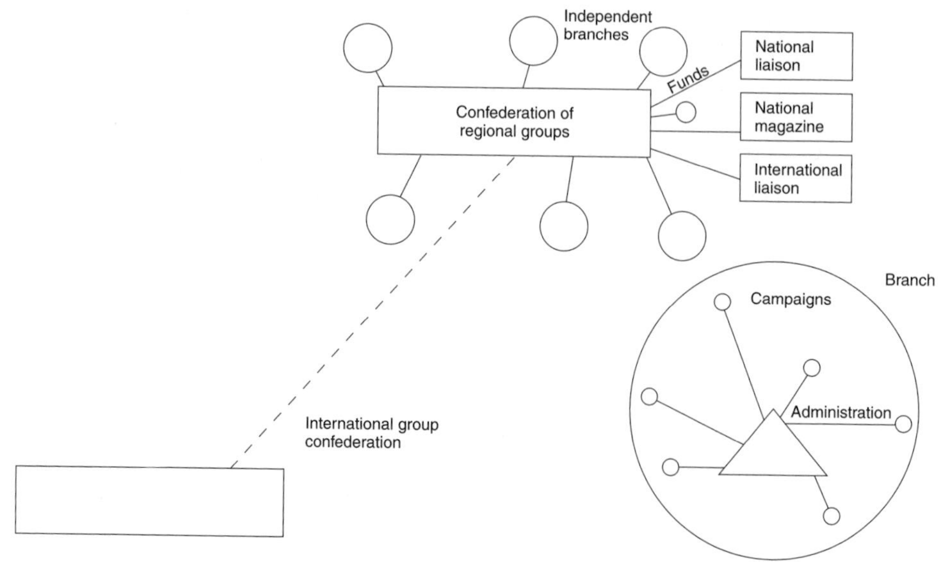

Figure 4.3 ***Friends of the Earth (structure)***

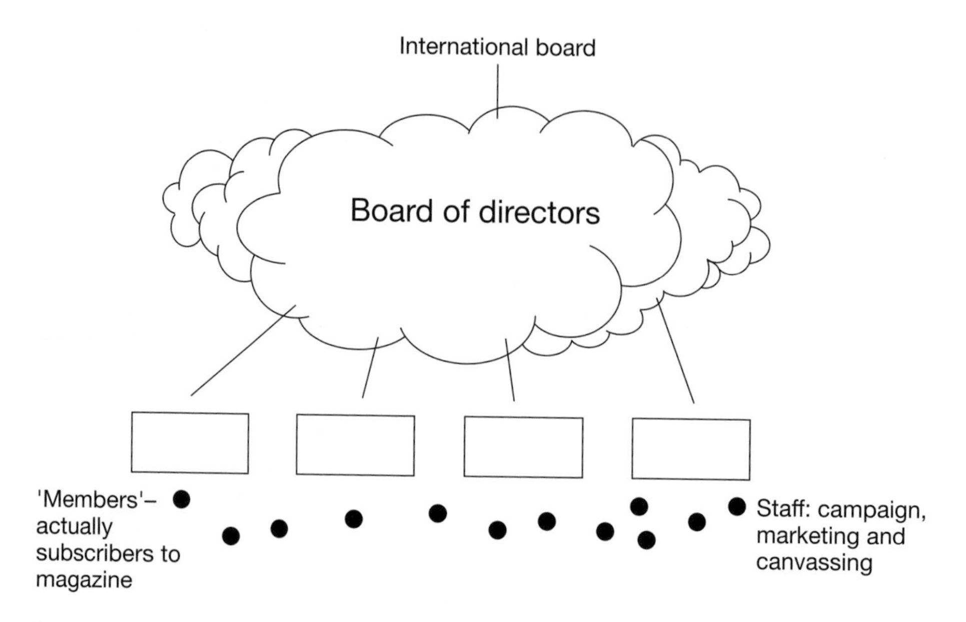

Figure 4.4 ***Greenpeace (structure)***

corporations. Additionally, there are often conflicts over levels of professional control versus the right of members to direct their own organisation. The case of the Sierra Club is most apt here. In the mid-1990s, a breakaway group emerged from within the Sierra Club called John Muir's Sierrans. This organisation was formed to 'reclaim the Sierra Club and restore Muir's (the founder of the Sierra Club) vision, passion and strength' (Hansen 1995: 26). Chad Hansen, a key mover in the society, writes about the Sierra Club management:

> At some point they lost sight of the fact that ecosystems are more important than the organization; that this is not about their organization, but a struggle with life and death consequences for millions of species and future generations of humans. They succumbed to the intoxications of political access – at a price. They lost their vision. They lost their connections to the land. After a while, there ceased to be a right and wrong – just access, ego, and power. While there's still hope – and there is still hope – it's time for this to change.
>
> (ibid.: 26)

Two points are raised by Hansen. First of all, green NGOs are rarely unified. Obviously, less conflict is tolerated in the more highly structured organisations than in informal networks but internal political and ideological conflict still remains regardless of the efforts to quell dissent. In the 2004 Presidential election, a major ideological faultline opened up in the Sierra Club yet again, but along very different issues. The Sierra Club's policy on restricted population growth was the site of this new contest, with powerful networks from the anti-abortion, right-to-life movement infiltrating the organisation before the election, in a bid to deliver green votes to the incumbent Republican Party.

The second point derived from Hansen's earlier comments relates to the pressures to formalise. Elite theorists like Robert Michels would say that oligarchy (rule by the powerful few) is the inevitable result of any organisation. He referred to it as the 'iron law of oligarchy': 'whoever says organisation, says oligarchy' (Michels quoted in Plamenatz 1973: 32). There can be no doubt that whether an NGO has grassroots origins or was 'born institutionalised', there are strong external pressures to mimic the politics of the *status quo*. But for every Sierra Club or ACF, there are also examples of organisations like Friends of the Earth that continue to promote the concept of direct democracy, despite pressures to the contrary. Rather than iron law, there is a tendency towards oligarchy.[9] Obviously, in those organisations that reproduce the political forms of

Plate 7 ***Indian environmental NGOs build transnational links at the World Social Forum. Courtesy of Joel Catchlove***

governments and/or corporations, these pressures are greatest. It is for this very reason that many of the more radical environmental groups and networks deliberately remain non-institutionalised. Formal organisation brings advantages and disadvantages to environmental politics. Increased short-term effectiveness and efficiency often come with the loss of long-term vision, and power becomes concentrated in the hands of a few.

Transnational green NGOs: Emancipation versus global green governance

As the previous chapter revealed, transnational environmentalism is on the increase in the new millennium. Green NGOs, as the most visible part of environmental movements, are also increasingly globalising their operations. Many organisations have responded to globalisation differently. In an excellent study of three British NGOs – WWF, FoE and Greenpeace – which have increasingly globalised their 'brand' over recent years – Rootes argues:

> all have transnational origins, affiliations and aspirations, but whereas WWF and FoE have broadened their agenda to embrace the concerns of the global South with sustainable development and social justice,

> Greenpeace appears little changed . . . Whereas Greenpeace has mainly worked alone on strictly environmental issues, WWF has engaged partners in government and corporations and joined campaign coalitions on global issues. FoE has eschewed corporate partnerships, but has played a leading role in campaign coalitions within and beyond the environmental movement, including the emergent environmental justice movement. Attentive to the concerns of its local groups and prominent in FoE International, FoE is uniquely exposed to grassroots pressures and influences from the South.
>
> (Rootes 2006)

Doherty and Doyle (2006) refer to two types of transnational NGOs in the global green public sphere. There are those 'emancipatory' NGOs which focus on building regional and global networks in a manner which increases the power resources of the poor and the environmentally degraded. In Rootes' schema, FoE is an excellent example of this 'ideal type'. These often construct themselves as separate from any notion of the state, and what they perceive to be a global neo-liberal project, working to break down these very structures of 'repression' through using the non-institutional forms of organisation described previously. The other type are particularly powerful and well-resourced environmental NGOs, which position themselves as part of the neo-liberal project of the global governance state. These 'governance groups' are those that offer little challenge to environmental injustice and are in general reproducing forms of inequality through their participation with governments, financial institutions and transnational corporations. They argue that only by engaging with the subjective voices of the local, traditional and indigenous peoples, can adequate ecological management strategies be assembled. They write:

> using limited – usually post-materialist – interpretations of green concerns to continue to discipline societies which do not mirror their own constructions of *nature*, or what, in their minds, constitutes a productive and *democratic* civil society (p. 2).

Again, if we refer to Rootes' study, WWF can be used as an example of this type of green NGO (also see Box 4.4 for other examples). However, even emancipatory NGOs operating at the transnational level can and have been accused of not going far enough in addressing inequalities, and thus becoming part of the problem. One such organisation is Friends of the Earth International, which is currently wrestling with recent frictions between some of their Northern and Southern branches (see Box 4.6).

Box 4.6

Tensions in transnational 'emancipatory' NGOs: FoE International and FoE Ecuador

Friends of the Earth International is a loose federation of over 70 national FoE groups with very different histories, resources and national contexts. Unlike the other major green NGOs, each national group has an equal vote at international meetings irrespective of the size of its membership or financial contribution to the federation. As the FoEI network has expanded, it has increasingly been driven by the major concerns of its Southern members, despite the wealthiest and largest groups being from Northern countries. FoEI has sought to address environmental issues through a 'critique of social and political inequality, an explicit attack on neo-liberalism and a commitment to environmental justice as a central principle' (Doherty 2006: 2).

Yet FoEI has faced a recent crisis due to divisions over North–South questions between different national groups. Tensions emerged during and after the second UN World Summit on Sustainable Development (WSSD) in South Africa in 2002, and led to the resignation of Acción Ecológica (FoE Ecuador, AE) who argued that Northern and Southern groups had contradictory ideological positions.

Ideological tensions were verbalised over several international campaigns. The corporate accountability campaign was criticised by AE who saw corporations as beyond accountability and argued that seeking to encourage reforms and regulation simply reinforces existing structures. It was felt that FoEI was driven by the search for positive media coverage which resulted in a weak expression of rejection of the business agenda at WSSD. The FoEI-claimed success of getting the World Bank's Extractive Industries Review to recommend pulling funding for oil and gas projects in developing countries, was also questioned by AE who felt it was an unsatisfactory outcome because it did not challenge the legitimacy of external investment per se.

Similarly, FoEI spoke only of achieving 'food security' when it was felt it should emphasise 'food sovereignty'. Another example was the use of the concept of 'ecological debt' by AE to challenge the reformism of some Northern groups. European FoE groups had been proud of their concept of 'ecological space' which was used to show the gross inequities of consumption patterns between North and South. But AE and other Southern groups felt it did not recognise the ecological debts imposed on the South by centuries of colonial exploitation, and they would only accept ecological space if it was accompanied by ecological debt. Ecological debt is a redistributive principle which involves redress for cultural and social injustice and thus draws on principles of recognition of past wrongs.

Finally, until recently, Southern groups were critical of the significance attached to climate change issues by Northern branches of FoEI. It was felt that climate change had emerged as an issue through Western science, and seemed to displace the current environmental injustices being experienced by a focus on a future problem. It was also asked whether desertification and deforestation were so comparatively low in priority because their main effects were in the South. Yet FoEI groups were able to work through their differences and agreed to place climate change campaigns against oil, gas and coal industries within the framework of climate justice.

The tensions expressed after the WSSD were dealt with by a Network Process Team created by the FoEI executive, a collection of eight representatives from the four regions who volunteered to examine how FoEI could manage its 'political diversity'. As part of the reconciliation process, by the time of the next international meeting in 2004 in Croatia, there was a willingness to admit to mistakes. Northern groups accepted that they had worked too much within the terms of events such as WSSD, and discussion was exchanged on political differences based on contexts and identities. Peace was re-established within the network relatively easily, based on participatory processes of dialogue to rebuild trust.

As a consequence, any claim about the identity or strategy shared by the FoE federation is now recognised as potentially difficult. While being the most inclusive international organisational structure, there are always inequalities of power within the network that affect its inclusiveness. However the process of tension and debate within the network was proof of the openness and resilience of the organisation and its ability to reflect on internal dissent, which in turn has assisted the constant evolution of the federation.

Source: Doherty, B. (2006) 'FoEI: Negotiating a Transnational Identity', *Beyond Borders*.

There is dramatic evidence that Southern NGOs are increasingly driving the global green agenda, with many Northern NGOs gradually moving away from a traditionally narrow conservation focus to incorporate the sustainable development and social justice concerns of the global South (Rootes 2006, Doherty 2006). Whether this is a 'radicalisation' of goals to reflect those of the South, or simply a means to co-opt Southern agendas is open to debate; but there is no doubt that this broadening of previously narrow conservation objectives in many parts of the North is a hallmark of transnational environmental NGO politics in an increasingly globalised world.

While some environmental organisations will seek to globalise environmentalism through disciplining the local into a carefully constructed and restricted version of the global (which can only be understood within a frame of post-colonialism no matter how honourable their intentions), the majority of environmental organisations across the world persist in forging resilient social alternatives for environmental emancipation.

Conclusion

To summarise, there has been an explosion in the number of NGOs dedicated to environmental concerns in the last three decades. NGOs are organisations and, as such, share a collective political form with other organisations, based on the existence of a constitutional charter. Despite this shared reality, they are still quite diverse in their characteristics: their geopolitical origins; their ideology; their size; the sphere of their political activity; their funding sources; and, most significantly, they differ in what tasks they actually perform. Finally, their internal structures (often informed by their goals and ideology) dictate the extent to which they establish relationships with governments and/or business corporations.

In recent times, NGOs have taken on an increasingly important role. With the emergence of global ecology, many environmental issues are seen as beyond the traditional scope of national governments. There have been attempts by governments to respond to global environmental problems through ventures like the United Nations' Rio Earth Summit and subsequent conferences on human populations and habitats. But governments are still lagging behind in their responses, and this transnational political space has been occupied by corporations and NGOs, which can cross nation-state boundaries more readily. This globalisation of ecological and market systems has led to 'the politics of no fixed address' (Doyle 1996).

Whereas traditional interest groups sought to influence those in government indirectly, green activists, some of whom have operated within NGOs, have decided to seek more direct entry into established forms of institutional power, including the formation of green political parties and taking part in elections.

Further reading

Bryant, R. and Bailey, S. (1997) *Third World Political Ecology*, Routledge, London.

Doherty, B. and Doyle, T. (2007) *Beyond Borders: Environmental Movements and Transnational Politics*, Routledge, London.

Hirsch, P. and Warren, C. (1998) *The Politics of Environment in Southeast Asia*, Routledge, London.

Lowe, P. and Goyder, J. (1983, 2004) *Environmental Groups in Politics*, Allen & Unwin, London and Boston.

Princen, T. and Finger, M. (1994) *Environmental NGOs in World Politics: Linking the Local to the Global*, Routledge, London and New York.

5 Political parties and the environment

- Electoral systems and politics
- Political parties
- Green parties
- Electoral success and failure

Introduction

> Political parties provide the primary method of selecting political elites and largely determine the content of electoral and legislative agendas; in addition, the parliamentary structure of most European governments converts partisan majorities into control of the basic institutions of governance.
>
> (Dalton 1994: 213)

Russell Dalton's comments are on events in Western Europe, where the most successful green political parties are to be found.[1] Political parties are central to the working of Western liberal democratic representative systems. The act of governing is driven by the actions or demands of the major political parties. Government in these systems is largely the result of party confrontation: 'It is party confrontation – automatic, blind, occasionally senseless confrontation, regardless of the nature, purpose or importance of the issue' (Jaensch 1983: 215).

In the past, electoral responses which included environmental concerns have been almost exclusively a Western phenomenon, in multi-party states. This largely remains the case, and this fact is reflected in the content of this chapter. In the new millennium, however, there have been three important developments: (1) a party-political response has emerged in non-Western democracies, such as is evidenced by the fascinating new green party experiences in places such as, for example, Taiwan and South

Korea; (2) for the first time, we have also seen one-party states – such as Cuba, China and Iran – having to confront and address the politics of the environment in serious fashion; and (3) even electoral politics is not immune from the processes of globalisation. In the past, political parties almost exclusively reflected the domestic political terrain within nation-states. The emergence of the Global Greens – a coalition of green parties around the world – challenges these preconceptions.

There are three basic ways in which environmentalists have responded to the electoral system: (1) by consciously abstaining from electoral politics; (2) by influencing existing political parties to take on elements of their ideological package (portrayed here as playing normal politics); and (3) by creating green parties. This chapter considers each of these positions in turn and concludes with an assessment of the relative costs and benefits of each of these choices.

Rejecting electoral politics

Many environmentalists – in both the North and the South – believe that formal, traditional politics is incapable of resolving ecological problems and partisan party politics exemplifies traditional politics. Party politicians are often perceived as having little interest in resolving difficult environmental problems or pursuing principled activities. Instead, they are seen as self-interested power seekers; strivers for power in a bid to create more power, as a means of staying in power. In places like Nigeria, Guatemala, Bolivia, Thailand, the Philippines or Fiji – where democratic regimes exist but are seemingly in constant transition to or from military control – many environmental activists believe that parliamentary politics is a corrupt domain, only paying lip-service to the concept of democracy. Instead of seeking change through electoral systems, environmental and social justice activists prefer to spend their limited time and resources on insurgency and mass mobilisation politics: resisting and mobilising against the power of the state, rather than playing its repressive electoral games.

In Western democracies in the North, other forms of criticisms are levelled at electoral politics. For example, even in liberal democracies which champion genuine electoral choice, often party platforms are produced which are profoundly limited in what they can deliver. The best way of attaining this power, this argument continues, is by providing a party platform that offends as few people as possible and certainly not those who hold significant power in society. Often major parties aspiring to win government begin to look more and more alike. Mayer suggests that this process is to be expected:

> It is not surprising that the two parties which make up the 'ins' and 'outs' should come to resemble each other markedly. Each, in striving for electoral support, tries to antagonise as few voters as possible and to win the support of as many as possible, and hence is driven to become moderate.
>
> (1969: 19)

If this is to be expected, then it should not be surprising that established political parties fail to challenge the *status quo* when it comes to environmental problems and that such parties make little effort to define and achieve green political goals.

Most environmentalists realise that there is little chance that greens will gain government in the foreseeable future (Germany and Sweden aside). So why play a political game that one cannot win? Even more apposite is the argument that ecological concerns are universal and must be accepted by all sides of politics. There is a fear that supporting one party will inevitably lead to a backlash, where the 'wronged' party will become hostile to environmental concern and good environmental practice. In addition, there is concern that not only the goals but also the manner of playing traditional party politics in traditional systems is adverse to the new politics of environmentalism.

Some of the problems of green groups playing party politics manifested themselves in the 2004 Australian federal election. A coalition of conservation groups in Tasmania endorsed the Labor Party's forest plan, which stated a commitment to end land clearing, protect old-growth forests and restructure the timber industry (TWS 2004a, 2004b). Labor promised that 240,000 hectares of high conservation old-growth forest identified by the Tasmania Together group would be the subject of a year-long independent expert scientific inquiry for the best conservation outcome (Thompson in Jones 2004). This Labor policy created tension between some elements of the Tasmanian Labor Party and the leader Mark Latham. The plan was also trumped at the last minute by an incumbent Liberal Party commitment based on a World Wide Fund for Nature plan to immediately place 170,000 hectares of forest into protected reserves. The fact that these 170,000 hectares were within inaccessible areas unwanted by the logging industry was not grasped by the general public, with the Liberals winning the election, in part, on the back of their forest policy. Environmental organisations like the WWF, believed that their involvement pushed forests onto the national agenda, leading to a positive 'greening' of the Liberal Party's policy on forests, despite the fact that overall, the Liberals had had a distinctly poorer environmental record over preceding years.

In an article strongly urging environmentalists to refrain from playing electoral politics, Brian Martin identifies seven points, which form the content of Box 5.1.[2] For these reasons and others, some French environmentalists argue: 'Elections: piège a cons' ['Elections: trap for idiots'] (Dalton 1994: 220).

It is interesting that Martin's position is largely based on Australian social movement experiences. Although involved in an active and, at times, successful movement, many Australian environmentalists have been extremely reticent about forming their own green party for many of the reasons Martin espouses.

Coupled with this reticence, there is a strong belief in the potential of the dispersed, connected, informal and longer-term politics of the social movement collective form. It was not until 1992 that a national green party was formed, and when compared with European green parties, it remains in its infancy.

Plate 8 ***Many of the grassroots members of the Greens were unhappy over Germany's entry into the Bosnian conflict and rallied all over Germany – like this one in Munich, Germany. Author's private collection***

Box 5.1

Reasons for environmentalists to refrain from electoral politics in liberal democracies

1. It does not challenge existing structures such as the bureaucratic organisation of the state or the profit system. Rather, entering election campaigns reaffirms the value of existing structures.
2. Focus on elections and dependence on sympathetic politicians does little to establish the social movement as a viable force outside the parliamentary arena. A basis for continuing struggle may not be established. Often after an exhausting election campaign, the movement virtually collapses.
3. The sense of personal responsibility for environmental problems is given away to elected elites.
4. Elected representatives, even those most responsive to community opinion, are still subjected to intense pressures to adopt anti-environmental or 'compromise' policies. Politicians are constantly influenced by industrial lobbyists and top state bureaucrats. A more pervasive influence is the requirement to maintain economic expansion in order to finance government programmes. Politicians cannot afford to jeopardise 'business confidence' by anti-capitalist policies. The key goal of political action becomes survival in office. For these reasons elites, including elected representatives, cannot be relied on to enforce environmentally sound policies.
5. Entering elections tends to polarise opinion on the environmental issue along party lines. Potential supporters in the party not endorsed become much harder to reach.
6. Election campaigning often depends on key personalities, either as candidates themselves or as charismatic campaigners. This dependence, plus the need to co-ordinate policies, maintain party unity and avoid doctrinal splits, tends to centralise power in the social movement itself, to reduce meaningful participation and thus weaken the base of the movement.
7. Strategies that do not depend on electioneering tend to be neglected.

Source: Adapted from Martin, B. (1984) 'Environmentalism and Electoralism', *The Ecologist*, 14, 3: 111

Sometimes, as touched upon, the decision to abstain from electoral politics can be a radical one. For example, those people who are involved in new social movement politics can see electoral politics associated with the formation of a green party as a distraction from their serious and difficult attempts at popular mobilisation for direct action. But such a decision to abstain – particularly in the North – may not be radical at all. Those who see their work as being primarily focused on pressure group activity may well view the formation of a green political party as an unwelcome complication. It is hard enough to try to persuade government to listen to environmental concerns without this being construed as partisan activity on behalf of a rival political party. There are also those who are members of existing political parties and who have worked long and hard to have their party add green concerns to its traditional agenda. Such people can see no reason for creating new political parties that make their task harder and challenge those to which they already belong. So it is quite possible to reject the formation of green parties to be involved in electoral politics on conservative, reformist or radical grounds, without in any way compromising a commitment to effective environmental action.

Greening mainstream political parties

As noted in the previous section, some environmental activists are involved in efforts to green the existing political parties. These people make a realistic assessment of the chances that any new political party will succeed in displacing the traditional parties and see their efforts best directed at getting their party to take green issues seriously. This is no easy task since green concerns have to jostle with a whole array of alternative claims about what is important. Nonetheless, in some circumstances, in some places, existing political parties have adopted some green issues as part of their political calculations about how to position themselves to win in conflict with their political rivals. In multi-party states, there are two ways in which political parties are 'greened': through the efforts of party activists; and through political calculation and party competition. There is also a 'greening' process in other types of governmental systems. Many political regimes of single-party states have also adopted some environmental concerns in their platforms. The process, of course, coincides with acceptance at ruling government level of the need for environmental protection.

Liberal democratic states

We will now turn to a few examples to understand the dynamics of the process within liberal democracies.

The Democratic Party in the United States has had an uneasy relationship with the politics of environmental concern. A number of its key figures and presidential aspirants have embraced environmental issues as part of their campaigns without a great deal of success. The image of Michael Dukakis being pilloried by George Bush's campaign ads for his poor environmental record, although full of irony, was none the less effective in marginalising the issue. In that successful presidential campaign, George Bush Snr adopted some claims of environmental concerns, but there was little evidence that this influenced his conduct of policy making. However, Bill Clinton, as presidential candidate, chose Al Gore, who was well known for his attempts to make the environment an electoral issue (Gore 1992), as his vice-presidential running mate. Undoubtedly, Gore worked hard to get the Clinton administration to adopt better environmental policies, and evidence can be found to indicate a degree of success. As soon as he was elected, Bill Clinton reversed George Bush's position on the protocols and communiqués issued at the Rio Earth Summit. Nevertheless, Clinton's response to the increased anti-regulation militancy of the Republican majority in Congress saw a reversal on several important environmental issues, including restrictions on the destruction of wetlands by private property development. During the 1996 election campaign, Clinton manoeuvred between green initiatives and lifting regulatory restrictions: a political calculation determining the extent to which his politics could be described as having been 'greened'. With the election of George W. Bush to the presidency in 2001, and re-election in 2005, this incremental 'greening' has been whittled away. Through its rejection of the Kyoto Protocol on climate change, the current US government is now seen as returning to a stance directly antagonistic to environmental concern.

A very clear example of how political calculation and party competition can 'green' an established political party is to be found in Australia during the years of Labor government (1983–1996). There was an extended history of party competition and co-operation over environmental issues in Australia. For example, the attempt to sand-mine Fraser Island was opposed by the Labor government of Gough Whitlam (1972–1975), but the decision to ban mining was taken after a protracted public inquiry by the Liberal government of Malcolm Fraser

(1975–1983). A public difference between the two major parties was a minor issue in the 1983 federal election, which saw a Labor Party success. At issue here was the willingness of Labor and the reluctance of the Liberals to use federal powers to override the Tasmanian state government on the issue of building a dam in the southwest rainforest 'wilderness' area of the state. So there was some important background for what happened in the mid to late 1980s.

The key to Labor's initial electoral success turned on its relationship with the trade unions and its ability to deliver sustained real wage cuts, undermining, to a degree, its electoral appeal. In these circumstances, Senator Graham Richardson took the environment portfolio and turned it into a significant force for environmental initiatives and a very effective part of a strategy to have the Labor Party re-elected (Richardson 1994). Richardson was the 'numbers man' for the New South Wales right, the dominant faction in the parliamentary party, a self-proclaimed realist, willing to do 'whatever it takes' to ensure Labor's success. What he did was to devise a 'green preference' strategy that would see those with strong environmental views vote in such a way that their preference votes would be transferred to Labor. His strategy involved making well-publicised, at times dramatic, important decisions to protect ecologically sensitive areas from economic development. This produced hostility in the business community and opposition from the Liberal Party but made it obvious that a Labor government would be better for the environment than its conservative opponents. In the 1987 and the 1990 elections, this strategy was a major part of the Labor campaign and its success. The strategy worked only up to a point, since the government did not adopt more than a 'green tinge' for its general policy making. Later, particularly with the arrival and ascendancy of the conservative Liberal/National Party government, the priority of economic development was reasserted and environmental support either declined or was neutralised, such that in the 1996, 2000 and 2004 federal elections several environmental organisations were confused about the relative merits of the two parties.

These two examples show both the scope and limits of any strategy for greening existing and mainstream political parties. In certain circumstances, these parties will adopt more green images and environmental concerns, but the extent of this commitment will always be limited by political and electoral calculations. When the situation is propitious or demands it, these green moves will be dropped in favour of other policy initiatives and other views on how to win elections.

One party states

Political parties adopting green concerns are not restricted to open democratic regimes, with many one-party authoritarian states also implementing green reforms through both choice and/or necessity since the 1990s. Cuba is probably the best example of this, where the end of assistance programmes from the ex-USSR necessitated adapting agriculture and energy policies to independently service the population, as well as responding to increased environmental concerns resulting from earlier rapid industrialisation (Lady 2004: 12). The Cuban Communist Party has had to find new ways to double its food production, halve imports, and maintain food export production to keep its economy above water; which it has achieved, to an extent, through an organic farming revolution. As well as in rural areas, thousands of co-operative gardens have been established by the government in urban empty lots, on rooftops, schools and workplaces. This has been combined with an educational programme to alert the public about environmental issues. Cuba now uses leftover sugar biomass to power nearly 30 per cent of its energy needs, and has sought to repair the damage done to its forests by involving half the population in tree-planting projects, and establishing national parks and protected regions across more than a quarter of a million acres on the island (Brodine, 1992: 23).

The ruling Chinese Communist Party has also become aware that failing to protect the environment incurs significant social and economic costs, and has been eager to reconcile both unimpeded economic growth and improved environmental protection (Economy 2004). An Environmental Protection Law was created in 1989; in 1994 the public were given the right to participate in environmental policy making, and in 2003 the government legislated public participation in environmental decisions, and the environmental impact assessment process (Tang 2004: 2). As touched upon, environmental NGOs have also begun to be permitted, such as 'Friends of Nature', opening some political space for popular participation in environmental protection (Economy 2004). While punishing dissent to government mega-projects such as the Three Gorges Dam, the Chinese government has slowly realised that environmental protection is needed and that people should be able to participate in that process.

The Iranian theocracy has also noticed a trail of environmental degradation, and appointed Iranian feminist Massomeh Ebtekar as head of the Department of Environment, where she particularly champions pollution issues (Doyle and Simpson 2006: 12). The state has developed new environmental laws, and has better educated lawmakers to uphold the few existing environmental laws such as the clean water and clean air acts. It has actively promoted green NGOs but, as aforesaid, all environment groups must be endorsed by the Iran Department of Environment prior to operation. The authoritarian theocracy allows the very slight greening of policy to the extent to which it does not pose any threat to the state. Even in one-party states, however, there are now cases where green protest parties (however illegal) have formed. Despite the attempts by the Iranian regime at restricting environmental concerns within their one-party electoral system, a dissenting Green Party of Iran now exists underground, as it were, with a website hosted in Canada. We now turn to these experiments in green party politics.

Creating green parties

Green political parties have been formed in many countries as a response to the experience of new social movement politics or the lobbying practices of environmental NGOs. A 'Global Greens' international network now operates to promote a Global Green Charter, deepen communication among Green Parties, and facilitate action on global environmental matters (Global Greens 2006). The most successful green party was formed in West Germany in 1980 and achieved electoral success at the national level in 1983. The radical style of the West German Greens, with their commitment to participatory democracy, leadership rotation and gender equality, contrasted sharply with the normal politics of the rest of West Germany's conventional parties. Post-reunification with East Germany, they formed a coalition government with the Social Democrats between 1998 and 2005, only losing a majority in government due to the losses their coalition partner sustained in the 2005 general election. Similar green electoral successes have now been experienced in Sweden between 2002 – 2006, when a minority Social Democratic government could only rule with the support of the Greens and leftist parties in exchange for their influence on government policy.

With the exception of the German and Swedish Greens (and the occasional state or local government party), however, environmental activists realise that it is very unlikely that they will be 'in' government, alone or in coalition. Why then are so many activists willing to give so much time and energy to the creation and running of green political

Box 5.2

Advantages of green parties in liberal democracies

1 To gain office as an elected representative.
2 To gain legitimacy, and to be seen as an 'authentic' political force by a wide audience outside of the movement.
3 To hold a balance of power.
4 To 'spoil'.
5 To make policy deals with more powerful parties.
6 To gain more access to the media. As the media is also dominated by electoral politics, environmentalists contributing to this sphere gain added legitimacy and their actions are more readily reported.
7 To count supporters, and more accurately judge the strength of opposition.
8 To gain financial rewards.
9 To lodge a 'protest' vote.

parties? Box 5.2 lists some of the advantages that can flow from green party involvement in electoral politics.

First, there is the possibility of gaining representation in an international, national, state or local representational forum. Such representation brings environmental concerns into parliaments, congresses and local councils. In this manner, environmental politics are either firmly placed on the mainstream political agenda or given wider publicity. This arrival of green issues on the electoral agenda signals the emergence of a new-found legitimacy. Environmentalists are now perceived as advocates of authentic concerns; no longer imagined by 'the public' as either peripheral or necessarily radical. This legitimacy may increase the popular appeal of environmental concerns and may attract more coverage from the mass media.

Enough green representation can give a minority party the balance of power in parliament and hence enhance its influence in bargaining over which party holds office and what kinds of policies are introduced. This situation existed for over a decade in the Australian Senate until the federal election of 2004, with Greens, Democrats and independents holding 'the balance of power'. Sometimes this balance leads a major party to make a post-electoral alliance with a green party in a bid to form government. This occurred in 1989 in the Australian state of Tasmania, where the Australian Labor Party could only form government by establishing an 'accord' with five green independents. Of course, from 1998 until the 2005 general election, the role of the Greens in coalition with the SPD in the federal government in Germany is an exemplar of what can be achieved.

Often green representation has not been so significant. Instead, policy deals are done with major parties before electoral contests. A good example is the 1996 US presidential election, where Ralph Nader ran on a green ticket in California. Due to the electoral college system, California is the most important state for presidential candidates 'to carry'. Nader knew that he could not win the election, but he believed that he could spoil the Democrats' chances of winning if enough disillusioned 'left-leaning' voters deserted the Democrat camp. The last thing the greens wanted was to see a Republican president, giving the Republicans 'a full house': the House of Representatives, the Senate and the presidency. Instead, Nader wished to use his campaign as a lever to get environmental concessions from President Clinton. There can be no doubt that after Nader decided to run, Clinton became more active in environmental reforms, as with his decision to oppose the 'salvage logging rider', which had increased logging in public forests since 1995. Nader repeated this strategy in the 2001 US presidential campaign. Many critics of Nader (including the outspoken Mike Moore) believed his green vote delivered the presidency to Republican George W. Bush, instead of Democrat Al Gore who had far better environmental credentials.

In both Britain and the United States, there is little prospect of greens gaining representation at the national level. Here, green parties use elections to attract media attention to their cause. In the 1989 European election, the British Greens, achieving a significant percentage of the vote, were separated out of the 'other parties' category in descriptive media analyses, and were highlighted as a fourth party next to the Conservatives, the Labour Party and the Liberal Democrats. This accomplishment saw a partial increase in media coverage of environmental issues, as the increased vote was interpreted by the media as a sign of extensive public support. The US Greens' 1996 presidential campaign was fought with a primary goal being 'to gain media coverage for their cause' (Kagin 1995: 24). There are also financial gains in playing party politics. Kramer writes about the West German context:

> They include reimbursement of election 'costs', public political 'education', political polling and research, political 'foundations', and so on. Of course only political 'parties' can join in this fun. The 5 per cent rule for parliamentary representation generally prevented smaller political formations from fully participating.
>
> (1994: 232)

The Green Party in British Columbia saw the possibilities of financial rewards. Party political contributions are tax-deductible in Canada, so it formed an Environmental Defenders' Fund with the slogan 'You pay $25 – Ottawa pays $75'. The money is used to support jailed direct-action forest environmentalists and their families by paying their fines and legal costs (Goldberg 1994: 13). Finally, minor parties can use elections as vehicles for protest. In Ireland, for example, many citizens vote for Comhaontas Glas in both local and European Union elections.[3] This support is not matched at national level. Many people feel that these secondary elections provide an outlet for a safe protest vote against *status quo* parties (Holmes and Kenny 1995: 224–5).

Why do green parties emerge and succeed in some countries and not in others? Most studies identify three types of factor to explain the rise and development of green parties. The first two relate to the level of environmental concern; and the availability of the 'right' political opportunities and contexts. Obviously, there must be some level of environmental consciousness in a given country if there is to be an electoral response. In countries of Eastern and Southern Europe, for example, where environmental concerns have been interpreted (particularly since the 'transition' in the case of the East) as a luxury concern, there is less chance of developing a viable party response than in other parts of Europe. More often than not, citizen support for environmental objectives only indicates a potential to be mobilised for electoral purposes. For example, Chris Rootes comments on European green parties in the following way:

> Two of the countries where environmental consciousness has been most consistently high and where the environment has ranked highly as a salient political issue – Denmark and the Netherlands – have produced only tiny and poorly supported Green parties, while in others where environmental awareness is less highly developed – such as Belgium and France – Green parties have been relatively successful. In Italy, where levels of both general concern about the national and global environment and 'personal complaint' about the state of the

> citizen's own environment were higher even than in Germany, the Greens have made only very modest electoral progress.
>
> (1995: 233)

Even a high level of environmental consciousness does not guarantee electoral success. Political opportunity structures must be favourable also. Being in the right place at the right time are critical determinants. Individual personalities also play a role, particularly in cultures that traditionally respect charismatic leaders. Sometimes, the chief factor that dictates success or failure may just be the standard of competency of party officials or the manner in which democratic needs are balanced by bureaucratic imperatives.

The previous sections discussed how some environmentalists worked with and within mainstream political parties to green their agendas and policies. In a country, state or locale where green concerns have been accommodated by established parties, there may be less scope for green parties to wave their alternative green banner; but if existing parties have failed to respond to environmental concerns, they may be more vulnerable to green electoral challenges.

Often electoral/political success may have little to do with the accommodation of green concerns, successful or otherwise. There may be, for example, a period of internal turmoil within one of the more powerful parties (sparked by ideological or strategic differences among its members), making that party less popular. Alternatively, it may be that economic or a multitude of other social scenarios have proved electorally unfavourable for a governing party. Whatever the reasons or non-reasons, mainstream parties are weaker during some elections, providing greater opportunities for green parties. On other occasions, the power of established parties may be overwhelming, shutting out any hope of minor party success.

The third and, without doubt, most crucial of these three factors determining electoral success, relates to the institutional structures and arrangements of the electoral systems themselves.

Institutional structures and arrangements: differing electoral systems

The rules of the political game, as defined in the constitutions, conventions or institutions of a particular polity, tend to determine what can or cannot be done in a given context. Often, the design of a political system is such that it is virtually impossible for new parties to enter the mainstream political stage. For example, it is virtually impossible for third parties to form and operate successfully in the USA. The laws in most states make it difficult for third parties to secure a place on the ballot by requiring large numbers of signatures, whereas the Democratic and Republican parties are often granted automatic access. Further, the public funding of campaigns is much more generous for the two major parties. At a national level, for instance, third-party presidential candidates receive money only after the election, if they have garnered more than 5 per cent of the vote, and only in proportion to their total vote. Major party candidates, by contrast, get large election grants immediately upon nomination (O'Connor and Sabato 1995: 461). Obviously, one way around the American system is to be a very wealthy person which is how a presidential candidate like billionaire businessman Ross Perot emerged in 1996. Even with his wealth he was unable to win, and hints at the substantial problems faced by any less well-funded green political party.

In Britain, Germany and Australia, it is far easier to register as a political party. In Britain's case only nominal deposits are required of parliamentary candidates, with 'no restrictions upon individuals offering themselves for election' (Rootes 1995: 238). Despite this ease of registration, the British Green Party has achieved little success at the national level. By far and away the most prohibitive institutional factor in dampening the British Green Party and even the American Green Party prospects is the nature of the electoral system in these countries. In addition, the relative success of the German Green Party and, to a lesser extent, the Australian Green Party is, in part, the result of a favourable electoral system.

There are three basic electoral systems that need to be considered: first-past-the-post, preferential and proportional. In the first-past-the-post or winner-takes-all systems of Britain and the United States, it is simply necessary to get the largest number of votes cast in an electorate to win. There is no parliamentary representation for second place getters and other minor parties.[4] First-past-the-post systems tend to produce governments based on a far larger percentage of seats held in parliament than the percentage of the popular votes cast. As a general rule, first-past-the-post works to entrench a two-party system and is believed to make for stable government.

The preferential system is used in the national and state lower house elections in Australia.[5] This system is not much more favourable for minor parties. A winning candidate must receive an absolute majority of votes (50 per cent plus 1) in a single-member constituency. But few parties' candidates receive an absolute majority on the first count, particularly in a country like Australia, where every citizen must vote. So, the preferences of the other candidates are redistributed among the higher polling parties until one party crosses the 50 per cent plus 1 threshold.

In this system, minority parties (including the greens) do have a greater say, however indirectly, as to who governs by directing their preferences. Often policy deals are forged between prospective parties realistically vying for government, and minor parties pushing for, in our case, more environmentally responsible legislation. In the Australian preferential system, minor parties have not won direct representation, but their preferences have been significant and in some elections the two major parties have made 'bids' for these. Indeed, the 1990 federal election was won by Labor, at least in part, on the basis of the flow of environmental preferences.

Proportional representation is the system most open to green political success and is used for elections in West Germany, Sweden, New Zealand, the Senate in Australia as well as in state elections in Tasmania. To a large extent, this explains the existence of green representation in these countries' parliaments. This system gives minority parties a far greater chance of being elected (Doyle and Kellow 1995: 132). Proportional representation systems vary greatly but they share certain basic elements. Instead of the 'winner-takes-all' equation of the first-past-the-post electoral system, seats in parliament are distributed on the basis of the proportion of the votes won above a certain minimum level. Mostly, in proportional representation models, there are multi-member constituencies, and all that is needed to be elected is a 'quota' of the votes – and the threshold is far lower than 50 per cent plus 1 of the preferential system. It is for this reason that defenders of the current British and American systems refer to their own systems as delivering 'strong government' and look with suspicion at proportional representation, arguing that it gives too much voice to minorities and favours coalition governments with the attendant prospect of instability. The scope for green success in proportional representation systems can inspire conservative political forces with an urge for electoral reform. This has been the case in the Australian island state of Tasmania. Tasmania's House of Assembly was elected on the basis of five electorates each having seven members. Hence, gaining a quota of 14 per cent of the vote in an electorate would lead to representation in parliament. On this basis green candidates were elected throughout the 1980s and, for a time, they held the balance of power. This unsettled the Liberal and Labor parties so much that in 1998 they combined to amend the Electoral Act of Tasmania and reduce the number of members in an electorate to five. This increased the quota needed for election to 16.5 per cent and was intended to end green representation in the Tasmanian parliament. In the subsequent election the greens lost three of their four members but Peg Putt was successful in winning a seat in Denison and maintaining a parliamentary presence for the greens.

The type of electoral system is the most pressing institutional factor dictating the existence and, to a large extent, the success of green parties in achieving representation in parliamentary systems. Rootes writes:

> In general, in countries with federal constitutions and proportional representation electoral systems, the institutional matrix is much more favourable for the development and success of Green parties, and for the development of mutually beneficial relationships between Green parties and the environmental movement, than it is in centralised unitary states with majoritarian electoral systems.
>
> (1995: 241)

Due to the over-riding importance of the institutional make-up of electoral systems, the remainder of this chapter will dedicate itself to providing national examples of these three multi-party electoral systems: proportional representation in Germany, Sweden and Taiwan, the first-past-the-post system of the United Kingdom and the United States; and the preferential system of Australia.

Proportional representation: the Green Party experiments of Germany, Sweden and Taiwan

No study of environmental politics is complete without visiting the phenomenon of the German Greens which tasted their first electoral successes as the West German Greens (Die Grünen). West German citizens possess a high level of environmental consciousness. Some theorists argue along the post-materialist lines outlined in Chapter 3: that 'a new educated middle class who have grown up under conditions of relative peace and prosperity' are the base from which German Greens derive their support. Others of a more post-industrialist bent argue that the unprecedented prosperity of West Germans led to unprecedented pollution, and this environmental degradation forced the West Germans' hand (Frankland 1995: 24–5). This is more closely aligned with the post-industrial arguments forwarded as explanation in Chapter 3. Whatever the origins of support, environmentalists in West Germany were intensely political. Not only was ecology a central plank, but so were participative democracy, social justice and non-violence. The environmental philosophy of Die Grünen, more than any other national experience, has shaped key tenets of global green electoral politics.

Germany's electoral system is dominated by proportional representation. Any party that achieves a 5 per cent threshold enters the national-level Bundestag. Germany is also a federal system comprising Landtages. As said before, federations also provide added advantages to minor parties. Consequently, these are 'more significant policy arenas than the local and regional councils of unitary systems' (Frankland 1995: 27). On the European Union stage, Germany's federated PR system is also readily apparent, giving the Greens greater access to far more direct electoral power than in any other of the three case studies. On 13 January 1980, a diverse alliance of activists launched the national West German Green Party: Die Grünen. 'They agreed that the pillars of the new party would be grassroots democracy, social concern, ecology and non-violence' (Frankland 1995: 24). Early successes in the Landtage were built upon with representation in the Bundestag in the 1980s. This began in 1983, when they polled 5.6 per cent of the national vote, which converted into 27 green deputies. Four years later, this representation grew to 42 deputies.

In the early 1980s, Die Grünen referred to itself as a 'movement-party', understanding that there were major advantages to be had if some of its social movement characteristics could be held on to and further creatively developed in conjunction with a partisan wing. Partly as a result, even today the German Greens have problems finding day-to-day operational activists, as many of its supporters insist on devoting large amounts of their time to extra-parliamentary activities (Poguntke 1992: 250). On a more positive note, the movement-party aspiration challenged the way party politics was being played in West Germany.

It held on to its grassroots aspirations and based its intra-party procedures on *basisdemokratie*: 'Grass-roots democracy means the increased realization of decentralized, direct democracy . . . We are determined to create a new type of party organization with decentralized structures that are designed according to the principles of grass-root democracy' (Die Grünen 1980, quoted in Poguntke 1992). Individual members and lower-level organisational units were given substantial power to participate directly in high-level decisions. The most famous example of these ideals was the 'rotation principle', which initially limited sitting MPs to two years in the Bundestag. This insistence on challenging the operational characteristics of the existing party framework led to many creative and sometimes destructive tensions within the party. Poguntke explains:

> In a nutshell, it can be argued that, organizationally, the Green Party represents the attempt to reconcile elite-challenging participatory aspirations with the constraints of party politics in a representative parliamentary system.
>
> (1992: 241)

This tension continued throughout the 1980s, often depicted as a conflict between two opposing factions: *Realo* versus *Fundi*. The Realos emphasised reform and playing party politics by the existing rules, including experimenting with alliances with the Social Democrats (SPD). The Fundis involved themselves in extra-parliamentary activities and supported more radical declarations (Frankland 1995: 32). Towards the end of the 1980s, Die Grünen began to suffer, in part because of this feud.

Even more crucial than internal machinations was the political scenario provided by unification. In the former East Germany, environmentalists have emerged from other roots. The Alliance 90 group, formed for electoral purposes by East German Greens in the 1990 first all-Germany election, have a style of environmental consciousness vastly distinct from their West German counterparts. Jahn explains:

> Their protest was against the SED (Socialist Unity Party of Germany, the former East German ruling party) state, and they stressed the importance of 'dialogical politics'. This attitude implies, in sharp contrast to the West German Greens, an objection to being given the image of a party of the left. It was to be foreseen that in some states (for instance Saxony) Alliance 90 would even enter into a coalition with the Conservative Party (CDU) and cooperate with the authoritarian Ecological Democratic Party (ODP). In the negotiation process with the West German Greens, Alliance 90 demanded acceptance of the political system of the Federal Republic of Germany and the free market economy.
>
> (1994: 314)

The 1990 general election was evidence that Die Grünen and the East German Alliance had not successfully merged. Die Grünen polled poorly in this spirit of national unification (losing all its Bundestag seats), its election slogan demonstrating how out of tune it was with the country's mainstream sentiments: 'Everyone is talking about Germany. We are talking about the weather' (Joppke and Markovits 1994: 235). Its allies in East Germany won eight seats. After this disastrous election, in May 1993, two parties formed into one: Alliance 90/the Greens (referred to as 'The Greens'). This alliance signalled the triumphant return of the Greens, polling 10.1 per cent of the national vote in the June 1994 European elections, giving them twelve seats in the parliament. This truly was a major victory. Jahn contends:

> The future of Germany and Alliance 90/The Greens depends on resolving the symbolic struggle between competing answers to the problems facing a unified Germany.
>
> (1994: 317)

Interestingly, this struggle remained connected, for a time, to the battle of wills between the Fundis and the Realos. In the late 1980s, it had seemed as if the more radical Fundis held the upper hand. But since unification, the pragmatic politics of the Realos has seemed more in tune with the imagined free-market, sustainable development futures whose virtues were being extolled in the East. Many of the Fundis have now left the party battleground, and the Realos dominate.

Also, the much larger West German membership dominates in post-reunification politics. What is most fascinating is a new willingness to forge alliances with major parties (particularly the SPD), and this has actually led to the greens *sharing government* in some Länder (states) and from 1998 to 2005 in the national government. This has led to some major environmental successes with Greens-initiated legislation, for example the 2000 'Nuclear Exit Law' which began the phase-out of all nuclear energy, and the subsequent Renewable Energies Act.

Actually being in government has led the greens into many new kinds of policy dilemmas and other issues. For example, in 1996 Joshka Fisher, a green member of the German parliament, voted to send troops to Bosnia as part of the Dayton Accord to protect the Bosnian Muslims: 'This has led to a split between pacifism and the prevention of genocide in the Green Party' (*New Perspectives Quarterly* 1996: 52–3). Similar tensions also emerged during the March 1999 bombing of Serbia. Air raids in which German pilots flying Tornado warplanes took part in the nation's first offensive military engagement since the Second World War caused 'endless unrest' in the pacifist sections of the Green Party (Schmid 1999: 2). Again in 2001, the party experienced a crisis as some Green members of parliament refused to back the government's plan for sending soldiers to help with the US attack on Afghanistan.

The Green Party in Sweden was founded in 1981, with its roots in the environmental movement, opposition to nuclear power, the women's movement and the peace movement. They first entered parliament in 1988 with 5.5 per cent of votes (Bettinga and Hollinger 2001), becoming the first new party in 70 years to do so (Miljöpartiert de gröna 2006). As part of the worldwide green political movement, their policies include:

> long-term sustainable, democratic societies where people assume responsibility, both locally and globally. [Their] vision is societies that live in peaceful coexistence, participate in equal collaboration and where humans, animals and nature are respected.
>
> (ibid.)

Like their German counterparts, they also advocate grassroots democracy, self-reliance and decentralisation; give greater weight to values and attitudes traditionally associated with women; and believe that involvement in politics should not be regarded as a life-long profession (ibid.). To this end they have no formalised leadership, preferring two party spokespeople instead, with one always female and the other always male to promote gender equality. As political competitors have adopted environmental issues, the Greens have realised that they cannot rely purely on environmental issues and so have also taken a strong stance against the 'United States of Europe' (the EU) (Lundstedt 2004: 7).

Sweden shares a similar proportional representation voting system to Germany. From the 2002 election until 2006, the Green Party shared government with the Social Democratic minority government and the Left Party. This was at times difficult since the Social Democrats opposed shorter working weeks; supported economic growth; wanted increased European cooperation and EU membership – whilst the Green Party opposed growth; demanded Sweden's withdrawal from the EU, and wanted to immediately dismantle existing nuclear power plants (ibid.: 6).

There was criticism that the Green Party had abandoned its traditional values and conformed to the will of the Social Democrat prime minister (Schlaug in Lundstedt 2004: 8), and that it had moved towards a relatively liberalist and centrist ideological profile during the last few years in government (Burchell and Williams in ibid.). Yet the Greens themselves admitted that there are disadvantages to being part of government:

> There is of course a risk that the ideology of a party becomes indistinct, which scarcely is in favour of our party. To prevent this, we have to continue being active, combining green politics with influencing public opinions. Radical demand concerning green changes is necessary, not only on the surface, but changes within the structures of the society as a whole.
>
> (Miljöpartiert de gröna in Lundstedt 2004: 9)

Outside of Europe, Taiwan is one country which also has a favourable voting system for smaller parties, but greens have had significant but relatively unsuccessful interactions with the mainstream political process to date. The green social movement in Taiwan, like Germany and Sweden, emerged from protest against nuclear power and nuclear weapons in the late 1970s, and in particular the building of a fourth nuclear reactor in Taiwan. However, the unique history of the Taiwanese anti-nuclear movement has led it to be overwhelmingly dependent on the Democratic Progressive Party (DPP) – the main political opposition to the Kuomintang (KMT) – and the implementation of the DPP's anti-nuclear platform once they gained power (Ho, M.-S. 2003). When the DPP were finally elected in 2000, they traded their nuclear bill, which would have halted construction of the fourth nuclear reactor, for other concessions from the KMT (ibid.: 702), in order to prevent a 'constitutional crisis'. This was to the extreme dismay and detriment of the anti-nuclear movement who had placed all their hopes on the DPP. It then also became very difficult to persuade any non-DPP politician to support the anti-nuclear cause, as endorsement was aligned with the DPP (ibid.: 700). Thus playing party politics became a real liability for the movement.

In an attempt to salvage the anti-nuclear movement from DPP dependence, a Taiwanese Green Party was formed in 1996. As a member of the Asia-Pacific Green Network, it too adopted the Global Green movement principles of ecological sustainability, grassroots democracy, social justice and world peace (Green Party Taiwan 2006). They have promised to be the true opposition party, and 'to replace the DPP as the political advocate for all the social movements' (Ho, M.-S. 2003: 703). Given the constraints of Taiwan's political system for green reform, they have decided to opt for a radical constitutional reform approach, with one platform being that the elected national assembly representative would donate 80 per cent of their salary to the movement for social justice and ecological sustainability, before the National Assembly would be abolished and a new green constitution implemented (Green Party Taiwan 2006).

Green Party Taiwan gained 2.97 per cent of votes in their first national election and won one government position, but has underperformed ever since. Today there are no elected officials or representatives, and local council elections have given them an average vote share of 1.1 per cent to 1.8 per cent (Ho, M.-S. 2003: 704). Ho attributes the Green Party's failure to the fact that it takes many years to build up electoral viability; that the anti-nuclear movement is so dependent on the DPP that it continues to support the ruling party rather than support a party with much less chance of winning; and the Green Party has never run a candidate against the DPP in the district of the fourth nuclear power plant where it would stand its greatest chance of success. The Taiwanese anti-nuclear movement is now no closer to its goals than it was at its inception and, according to Ho, the green cause has suffered as a result of failed political party involvement.

First-past-the-post systems: Britain and the USA

'Britain has the oldest, strongest, best-organised and most widely supported environmental lobby in the world' (McCormick 1991: 34, quoted in Richardson and Rootes 1995: 66). This claim, however debatable, has some grounds for support. British environmentalists, as described in Chapter 3, have a far greater range of environmental positions than those dominant in the Australian and US experience: from nature conservation to political ecology. Like many European environmentalists/ecologists, environmentalism in Britain is perceived by many as intensely political, and a sustained systemic critique is visible within and outside the movement. The most recent of these more radical political traditions emerged in the 'anti-roads' campaigns that challenged that industrial icon, the 'car' and promoted the alternative concepts of sustainable villages and cities (Wall 1999; Doherty *et al.* 2000; Doherty 2002; see Box 3.1 for synopsis of anti-roads movement).

Unlike most other parts of Europe, Britain also possesses a first-past-the-post system, designed for 'big government'. This, more than any other factor, limits green partisan success. Even in 'secondary' European Union elections it maintains this system.

The British Green Party history is a relatively long and chequered one by green standards. The Ecology Party was formed in 1973 by a small group of activists from Coventry (Bramwell 1994: 121). According to Richardson, the British experience was dominated by the ideologically 'purer' ecologists (those who attributed intrinsic value to the Earth) rather than the environmentalists (those who simply cared for 'the betterment of human society') (Richardson and Rootes 1995: 7).[6] In 1985, it was renamed the Green Party and now has numerous branches in both local and national electorates. The late 1980s to the mid-1990s were characterised by mixed results and internal politics (Hutchings 1994a: 20–1). The internal disputes were not unlike the more famous West German disputes between the fundamentalists and the realists. Rootes observes:

> The dispute within the party is not about matters of environmental policy but about organisational structure. 'Realists' like Jonathon Porritt and Sarah Parkin assert that the party has squandered public sympathy by giving an impression of confusion and disorganisation and argue for a formally elected leadership, while 'fundamentalists' regard all talk of leadership as anathema and argue for maximum decentralisation within the party.
>
> (1995: 85)

Nationally, the British Greens polled their greatest success in the 1989 European election with 14.9 per cent of the vote, the highest achieved by any Green Party in the EU. This was after a series of events had heightened public concern on green issues, such as 'a new awareness of global environmental problems such as the depletion of the atmospheric ozone layer and the "greenhouse effect" of carbon dioxide emissions' (Rootes 1995: 71). This was strengthened by more local concerns such as the quality of drinking water and the safety of basic foodstuffs. It should be noted that at this time the Conservative Party was emphatically dominant, that Labour had lost the national election badly and that the Liberal Democrats were not offering an effective alternative. A green vote was, at the very least, a very safe protest vote. Rudig and Franklin observe the following about this election in 1989:

> [the] special circumstances of the European elections appear to have favoured the Greens. Environmental issues had played an unusually important role in the preceding year, the center parties were in steep decline, and (more than any other EC country) European elections were not considered particularly important affairs; the political cost of voting Green was thus very low.
>
> (1992: 39)

There was some hope that these electoral gains would somehow translate into similar wins in the 1992 general election. This did not eventuate, and the Green Party polled only 2 per cent (approximately) of the vote. On one occasion, it suffered the humiliation of being beaten in a by-election by the Monster Raving Loony Party. Its political profile at both the local level and in the European Union is far more favourable. The Greens have had considerable success in local government. There are anywhere between 100 and 300 councillors scattered across England (Hutchings 1994b: 20–1), 80 per cent of whom sit on the bottom tier: the local (parish, town or community) council. In some rural areas, green parties are more conservative than some of their urban compatriots.

This may be attributed, in part, to its stance against the euro and in support of a system of locally sensitive, decentralised and democratic economies. The Green Party of England and Wales, like many other European Greens, has led a campaign in opposition to the euro (the 'NO' campaign of 2002). They advocate that a single European currency is enhancing the undemocratic and unsustainable effects of globalisation, and call for communities and local governments to 'take back control over their local economies and enable them to rebuild stability' (Green Party of England and Wales 2002).

Without reform of the electoral system, green parties will remain peripheral in British politics. This accounts for a number of central green figures being active in Charter 88, the campaign for constitutional reform (Richardson and Rootes 1995: 86). In Britain, at the national level, the 1989 European Union election remains the pinnacle, with the greens polling 14.9 per cent of the vote. Unfortunately, this did not translate into seats in the European Parliament.

A similar story can be told of the United States. In 1992, 70 to 80 per cent of Americans called themselves 'environmentalists' (Easterbrook 1995: 26). North Americans, in general, are similarly dominated by the post-materialist values of the environment. The 'environment' has been dominated by non-human wilderness concerns and debates about pollution.

National elections are dominated by first-past-the-post electoral processes in the United States. Coupled with the immense difficulties of initiating a minor political party, this constitutes a major hurdle to green parties forming and becoming successful.

US Democrats, also, are clearly seen by many environmentalists as far more appealing than the Republicans. Other environmentalists argue that the Democrats do not go far enough to achieve adequate environmental reforms. The vast majority, however, do not involve themselves in third-party politics but pursue change through lobbying and, sometimes, direct action.

Of the major global North countries, the US history of green electoral politics is the poorest. Several green parties have been formed at local and state levels. At the national level, the 'Green Politics Network' was formed at a conference in June 1995. 'It seeks to group, under one spreading redwood, radical environmentalists and feminists, multi-cultural leftists, campaigners against the military budget, and that dwindling if hardy faction of citizens who still call themselves "socialist"' (Kagin 1995: 24). This network helped to foster the move to run a presidential ticket in 1996.

There have been more numerous local victories, however, but few are better than those enjoyed by the New Mexico Green Party. In March 1994, greens landed a seat on the Santa Fe City Council. In part response to this success, Roberto Mondrano ran for governor, and Lorenzo Garcia ran for state treasurer. Mondrano won 11 per cent of the vote; while Garcia won a record-breaking 33 per cent of the vote: 'more than any third-party candidate to run for state office in the last sixty years' (Skinner 1995: 14). The ousted Democrat governor accused the greens of being 'spoilers'. The greens' showing in the election gave them major-party status, which guaranteed them a place on the ballot at the next general election, as well as giving them the right to hold primaries (Cooper 1994: 453).

The George W. Bush era, however, has seen the national US Green Party steadily lose popularity, especially at a presidential election level. The party is now in a state of disarray, after the 'heavy infiltration of Democratic Party operatives' meant that the popular Ralph Nader was dumped for the ticket of Pat LaMarche and David Cobb in the 2004 presidential election (St Clair 2004: 30). In that election the Green Party allegedly instructed its members to vote for its candidates 'only in states where their vote doesn't matter' (ibid.), with the belief that any president would be better than George W. Bush. The fact that the alternative, John Kerry, had similar policies on, for example, the war in Iraq, to Bush did not seem to matter. While the Green Party formed 'not in opposition to Republicans but from outrage at the rightward and irredeemable drift of the Democratic Party' (ibid.), it now acts as a 'private contractor' for the Democrats according to St Clair (ibid.). The Greens only polled 0.1 per cent of the popular vote, while Ralph Nader ran as an independent with genuinely alternative policies and received the third greatest number of votes with 0.38 per cent (still far short of the 2.73 per cent he received in the 2000 election) (Leip 2005). The poor 2004 showing is evidence of the entrenched two-party system that first-past-the-post creates and the difficulties that green alternatives face. They can either join that two-party system and support the 'least worst' party that has a chance of winning; or continue to stand firm in their policies, attracting protest votes which translate into electoral oblivion. Either way, the chance of green success is quite slim in such all-or-nothing systems.

Preferential system: Australia

A high percentage of Australians have been measured as having environmental sensibilities. Most notions of environmentalism are post-materialist, oriented around wilderness concerns. Three times more Australians are members of environmental organisations than political parties.

In Australia, there are preferential voting systems for both state and federal parliament 'lower houses' (except in Tasmania). This has not led to green parties achieving representation in the House of Representatives. In the 'upper houses' (including the national Senate) there is proportional representation. All states have upper houses of 'review', except Queensland.

Both the Australian Labor Party (ALP) and the Australian Democrats (AD) have sought to attract the green vote, in part delaying the establishment of a national green party until 1992.

Green parties have existed in Australia since the formation of the United Tasmania Group in the late 1970s. They exist at the local, state and national levels. The national Australian Green Party formed only in 1992; it was established so late for a variety of reasons. In addition to those listed above, there were intense rivalries between the national greens and smaller green parties in both South and Western Australia. Also, there is a long-standing sentiment among many Australian environmentalists against participating in more mainstream political forums, like political parties.

There has been recent, moderate success on the Australian scene. Two green candidates won election to the Senate in the 1996 federal election. Among these is the Green Party leader, Dr Bob Brown, made famous by his involvement in the Tasmanian Wilderness Society's successful bid to

Plate 9 ***Great Ravine, Franklin River, Tasmania. Courtesy of Chain Reaction***

stop the building of the Franklin Dam in Tasmania's southwest forests in 1983 (Doyle and Kellow 1995: 133). In 1989, also in Tasmania, five green independents were elected to parliament. They briefly formed a government by creating a Labor–Green Accord. During the mid-1990s, two West Australian green senators sometimes held the balance of power in the national upper house, and used this position to bargain successfully with parties both in government and in opposition.

During the decade of the Howard federal government (1996–present day) the Australian Green Party has operated quite successfully with representation in the upper house (both federal and state), which is elected under proportional representation. No green representative, however, has ever attained power in the preferentially elected lower house. With the election of a conservative/neo-liberal majority in both houses of parliament in 2004, the Green Party's ability to wield its 'balance of power' in a way which influences legislation has been lost. Still, it has continued to provide an alternative voice to the two major parties on issues as diverse as immigration policy, higher education and the deployment of troops in Iraq and Afghanistan.

Conclusion

There are three clearly different ways for environmentalists to play electoral politics. Choosing the first path, deliberately abstaining from party politics, promotes the possibility of longer-term, perhaps more profound, changes. This is the political pathway of the outsider. The second path leads to insider politics; to play with and within major political parties. Increased political access and the possibility of being part of a governing party may lead to favourable policy changes. The trade-off, of course, is that increased access is also achieved on behalf of such a political party into the affairs of environmental movements. This may lead either to genuine reform or to the co-option of environmental goals. The final option is to form separate green political parties. This pathway is different from the others and it leads to two alternatives. It is possible to become so immersed in the party system that a green party begins to operate in much the same way as other established parties, with top–down leadership, centralised and hierarchical control of policy development, and a need to dominate broader mass movements so as to co-ordinate their actions with the party's electoral goals. A green party can also become committed to 'lowest denominator policies' designed to appeal to as many citizens as possible rather than being based on obvious and green principles. The alternative is to manage party politics in a manner which recognises that the green party is just one useful part of environmental movements, an electoral wing, not the evolutionary end-point for environmental dissent. This final pathway recognises that the different parts of environmental movements, political parties included, provide alternative venues for different types of environmental politics. Not all these are equally appropriate in changing specific political terrains. The greater the variety of ways of being politically active, the greater the chance of being politically effective in all the changing circumstances that confront environmental movements. Too great an emphasis on green party political activity may reduce the overall effectiveness of environmental movements.

Further reading

Martin, B. (1984) 'Environmentalism and Electoralism', *The Ecologist*, 14, 3: 110–18.

Richardson, D. and Rootes, C. (1995) *The Green Challenge: the Development of Green Parties in Europe*, Routledge, London and New York.

Wall, D. (1999) *Earth First! And the Anti-Roads Movement: Radical Environmentalism and Comparative Social Movements*, Routledge, London.

6 Business politics and the environment

- Business and the state
- Rejection, accommodation and environmental business
- Transnational corporations
- Managerialism and education

Introduction

Private business is one of the most important forces shaping the interactions between humans and the environment. This is not to say that private business is the sole force damaging the environment. Nor is it to assert that state-owned business has a more benign impact on the environment. Events in the Soviet Union and Eastern Europe should be sufficient to undermine such an interpretation (Carter and Turnock 1993). Nonetheless, private firms going about their normal business, unrestrained by too much government regulation, have delivered very significant amounts of environmental damage. Business has also been willing to invest considerable time and energy in resisting environmental regulations in order to protect its ability to continue doing whatever it wants without interference. As such, its political actions in defence of its ability to harm the environment need to be considered. But not all firms are indifferent to environmental damage. Parts of the business sector have worked tirelessly to develop and sell technologies that counter the environmentally harmful effects of industrial practices (companies developing pollution-control devices fall into this category). Some firms choose to work with technologies that are inherently less harmful. Some have redesigned their production processes to reduce the adverse consequences. Others have 'gone green' in the marketing of their products. These variations in responses from businesses require some explanation.

This chapter starts by considering the power of business and the various political positions that different firms take towards environmental care and concern. It then considers the different ways in which business has responded to environmental criticism and the various arguments and strategies used by business to defend itself against both criticism and regulation. Included here is a discussion of business attitudes to sustainable development and the belief that increased economic and technological efficiency is a way of limiting the harm done to the environment. The chapter concludes with a set of case studies to illustrate the scope for business to take different positions on environmental questions, in different contexts, and with different management structures or philosophies.

Business power, environment and the state

What is business? Business is a way of organising economic activity (either in the production or circulation of goods and services) whereby private ownership of the means of production is combined with labour so that the resulting output can be sold to others to realise a profit. Characteristically, business is organised into firms or corporations that have a given legal form. This form both creates them as legal entities and provides protection against various kinds of legal action. For example, the limited liability company (which separates the owners and managers of a company from its corporate existence) was a very important invention that made it possible for entities to grow larger in size and more ambitious in action than was practicable for individual or family concerns. Companies and company structures may vary greatly in size and kind, depending on the economic and legal circumstances in which they operate. The basic goal of all private business is profit. If the money invested in a firm does not return a profit, its creditors and shareholders cannot be paid and the firm will cease to operate. In this case, the firm will either be declared bankrupt or will be taken over by another company. Private business lives in a world of potential or real competition. Individual firms confront rivals, and the presence of competition will force firms to seek to contain costs while maintaining profitability. It is certainly true that the size of major companies has increased greatly over the years and that, in some parts of the world (and some sectors of the economy), a particular firm may hold a monopoly (or be part of an oligopoly) that allows it to dominate a market. Monopolies and oligopolies certainly help to reduce the pressure of competition but they do not remove it completely. There is always the threat of a rival entering the sector, or of a technological change that undermines a dominant position. The scale of major companies has also changed as national economies have become increasingly global. Imperial expansion encouraged big firms to operate overseas, expanding their investment in international production and trade. Gradually these large firms have assumed a multinational or transnational character, linking stages of production and circulation on a global scale across national boundaries. For example, a sports shoe firm may have its headquarters and image firmly located in the USA, have production facilities in Korea and the Philippines, use materials produced in a whole range of other countries, and be selling the shoes in both suburban and outback Australia. The scale of operations encourages flexibility and grants firms the ability to switch between countries on the basis of any marginal advantage. All these factors have consequences for the way in which business deals with the environment and responds to both criticism and the prospects of environmental regulation.

Private business, given the pre-eminent role it plays in the economy, has numerous resources that it can use to protect its actions from criticism. Threats to reduce employment, investment or the ability to earn foreign exchange can all be used as power resources in battles over environmental concern. For example, if environmental critics seek to stop a mining project, the business concerned can always point to the project's economic benefits. These would include the number of people who would be employed in the construction and operation of the mine; the amount of money to be invested in the project; the commercial flow-on for other firms in the area (such as local suppliers); the value of the ore when processed or exported; and, finally, the amount of tax or royalties that would flow to the various levels of government. These are all significant economic contributions to a local area and a nation, so the firm's case can routinely be expected to be very persuasive. Indeed, so strong is the presumption in favour of private business and economic development that it will be relatively rare for business to deploy its power resources in obvious pro-development campaigns. In normal circumstances, the implicit power of business is sufficient to allow it to proceed with business as usual.

When considering the potential and actual power of business, it is important to consider the contrast with most environmental organisations. If the clash is between a large firm and a local environmental group (even if that group has links with national or international environmental organisations), then the business will almost always have more power resources to deploy than the environmental activists. More than that, the power resources to be deployed will be different, or at least skewed in distribution. This power inequality is something that the media rarely highlights (see Box 6.1). In addition to the 'business' resources already noted, large firms have good quantities of money and professional resources that can be deployed in media and legal campaigns. By contrast, environmental organisations have to rely on their members for time, money and enthusiasm to mount and sustain any campaign. For environmentalists, co-operation and their ability to organise are important for offsetting these intrinsic sources of weakness. To be effective, organisation is needed to make a campaign work. For business, especially large businesses, organisation is an option rather than a necessity. Organisation here means a number of firms combining together to set up a specialised body to lobby government on their behalf and to co-ordinate a public response to, among other things, environmental challenges. Small firms are more vulnerable in any conflict (economic or political) and need greater organisation to be effective, even though competition makes it hard for them to organise in their own right.

The different role of organisation and the unequal distribution of power resources are not the only factors that govern what business has to do to protect itself from critics. As discussed in Chapter 1, the political regime is very important in determining what happens in environmental conflicts. Consider the bald difference between authoritarian/military rule and a more open democratic representative setting. The international oil company Shell had a relatively easy time defending itself against its Ogoni environmental opponents in Nigeria in the mid-1990s. Shell was so important to the Nigerian economy, and the revenue requirements of the ruling military regime, that there was very little doubt that the regime

Box 6.1

The business of the media in liberal democracies

As well as being big business, the mass media provide an important arena for conflicts arising from competing resource use and conservation values. While claiming neutrality of bias, a whole series of structural and instrumental influences shape media depictions and make the media yet another force in determining environmental outcomes.

The mass media does not usually accurately represent the nature of social movements – the amorphous, decentralised structure which includes both people, organisations and ideas. The mass media reflects dominant politics, and in particular, party politics, and sees the pluralist model as political reality. The pluralist model places party and pressure group politics at the centre of its analysis of 'normal' politics, while underrating and undermining the politics of popular social movements. Thus environmental movement initiatives are seen as 'single issue', 'interest groups' which come into conflict with other 'competing interest groups', and must lobby or pressure major political parties to be given any legitimacy by the media.

This has a flow-on effect in media reporting whereby movement participants are seen as an 'abnormal' sectional interest, rather than the average person. This leads to the media's use of stereotypical deviant images of environmentalists, both to denigrate and then self-perpetuate these images. The media stigmatises those who engage in environmental action in everything from how they look, to what they eat, to how they act, to their age or wealth. Generally those people who hold less power in society – the young, the old, the poor – and others who are not socially acceptable, are the most targeted. There is also a familiar theme of environmentalists being described as outsiders, and not legitimate community members (even if they are). Within this restrictive model, 'the media upholds the status quo, amplifies deviance with the purpose of stamping it out, and is the judge and the jury of our societal morality' (p.165).

The media is only able to sell personalities and stereotypes, which can at times aid the environmental movement as it did the anti-roads movement in 1997, but is a double-edged sword (Wall 1999: 91–2). In this campaign, the media 'celebrated roads protest and crystallised the actions of thousands of grassroot campaigners into the form of a single personality' (p. 91). This young deified tunneller 'Swampy' (Daniel Hooper) became a household icon and was briefly more famous than many politicians. However Hooper had not asked for fame and was embarrassed by his role, and direct action was transformed into a media spectacle – 'something for heroes rather than adults and children in local communities' (p. 92). The rejection of personality was seen as a means of making protest accessible rather than something 'unique' to a few individuals, which was lost.

In this the age of the media spectacle, most environment groups now deliberately target the media through 'direct communication' actions as the only way of gaining coverage – a means rather than the event itself being an end.

Source: Adapted from Doyle, T. and Kellow, A. (1995) 'The Media and the Environment', Chapter 7 of *Environmental Politics and Policy Making in Australia*, Macmillan, Melbourne, pp. 158–74; and Wall, D. (1999) 'Twyford, Other Tribes and Other Trials', *Earth First! and the Anti-Roads Movement: Radical Environmentalism and Comparative Social Movements*, Routledge, London, pp. 91–2

would stand by Shell and give it the protection it required to continue operating. In exchange, Shell would not interfere or make comment about the use and abuse of power by the ruling junta. This explained, in large part, why Shell maintained silence when the 'Ogoni Nine' were hanged in 1996. Since the subsequent advent of a democratically elected regime in Nigeria, Shell's relatively free rein has been checked somewhat: in 2005, the Nigerian court system found Shell guilty of negligence towards the Ogoni people and their lands, fining Shell damages worth approximately US1.4 billion dollars. Despite Shell's prior publicised respect in abiding by the laws of the military regime of the 1990s, it has refused to abide by the laws of a democratically elected Nigeria, as – at this time of writing – it has refused to pay the damages awarded by the courts to the Ogoni people.

On the other hand, Shell operating in Western Europe is much more vulnerable when it is strongly supported by the relevant governments. Consider the political problem the company faced when it wanted to sink its Brent Spar oil rig in the North Sea. Greenpeace organised a well-publicised occupation of the rig and a strong media campaign presenting the case that it would be ecologically irresponsible to proceed with at-sea dumping. The British government, as mindful of the importance of Shell as the military regime in Nigeria, tried to support Shell's plans and give the proposed sinking good environmental credentials. Such government support continued as environmental protests mounted. In the end, Shell gave in and abandoned its cheap method of disposal. In the British/European case, the nature of the political process, its openness in particular, made it very difficult for even a powerful company like Shell to prevail. International protest put pressure on Shell, but it could not save the life of the Ogoni activist Ken Saro Wiwa. This suggests that some political regimes are more effective in their support for business than others.

Lest it be thought that the above contrast means that international business would always prefer to work with an authoritarian rather than a more democratic regime, it should be noted that authoritarian regimes can also cause problems for business. Shell in Nigeria in the mid-1990s was not free to do what it liked. It needed to seek the favours of the military regime for its continued presence in the country and mute any criticism it might want to make so as not to offend military figures. None of this explains or justifies the poor environmental management of Shell's Nigerian operations or the environmental damage it continues to do.

It is important to recognise that authoritarian regimes and liberal-democratic representative regimes provide different settings and pose different challenges for business when confronting environmental critics. Authoritarian regimes tend to make life easier because such regimes are more likely to repress or restrict the actions of environmentalists, especially when they challenge development

Plate 10 ***Korean trade unionists on the final day of the World Social Forum, when tens of thousands of participants took to the streets through downtown Mumbai, India. Courtesy of Joel Catchlove***

projects that are close to the hearts and pockets of the rulers. Liberal-democratic representative regimes provide much more scope for environmental activists and create much more authority for governments to intervene against specific proposals while maintaining support for private business as a whole.

Sometimes it is not only the type of regime which is important, but rather whether a particular country is first world or third world. Consider the case of gold mining on the Ok Tedi River in Papua New Guinea (PNG). PNG is a liberal-democracy; but it is a very poor country shaped by successive colonial regimes from Germany, Britain and Australia (Hyndman 1991). Its attempts to secure economic development have been conditioned by an urgent need to win foreign exchange, which comes most easily from mining and forestry. Mining ventures have proved to be quite risky given the natural and social conditions in which mining takes place. A large copper mine on Bougainville Island (run by CRA) produced major environmental damage and social disruption, which culminated in the closure of the mine when an independence movement launched a war against the mining company and the PNG authorities.

In other areas, high rainfall and remoteness combined to make mining difficult. It was in such circumstances that the PNG government sought to lock foreign investors into a mining project in the relatively remote area of the Star Mountains at Ok Tedi. High rainfall and unstable geological conditions combined to make the prospective copper and gold mine quite challenging. After trying a number of partners, Broken Hill Proprietary (BHP, Australia's largest mining company) came to own and manage the project through OTML (Ok Tedi Mine Ltd – a holding company in some kind of partnership with the PNG government). The economic importance of the mine is very great, providing a significant portion of the export earnings of PNG, a sizeable amount of revenue for the PNG government, and feeding economic and social development in the mine region along the upper Fly River. Along with this economic benefit has gone considerable environmental damage and social dislocation which would not be tolerable in an affluent world country. The mine operates without a tailings dam, which would normally capture a lot of its waste. At one stage, there was an attempt to build such a dam, but it collapsed and has never been replaced. As a result, the tailings, crushed rock and eroded topsoil simply wash into the Ok Tedi River, increasing the sediment load. This leads to flooding and reduced fish supplies in parts of the river system. There has been a complex interplay between local land owners demanding compensation for the environmental damage and the PNG government's efforts to protect its project. BHP has expended considerable energy in defending its environmental record with claims about the naturally turbid condition of the river and the economic benefits of the mine (McEachern 1995).

From the perspective of the discussion in this chapter, what is important is the contrast between the operation of BHP in Australia with one set of environmental regulations and its operations in PNG, where weak government basically gives BHP a free hand to manage and monitor its own environmental impact. Mines have been run in Australia in ways that significantly damage the environment. For example, the Mount Lyell copper mine at Queenstown in Tasmania roasted its sulphur-rich ore with little regard for the consequent acid rain. As a result, the area around the mine resembles a barren 'moonscape', with little or no vegetation. This mine also operated without a tailings dam until 1994, so its waste polluted a significant area of Macquarie Harbour. Despite this record, it is not conceivable, even in the circumstances of a backlash against environmental care, that a new mine could be opened by BHP in Australia without a tailings dam to protect local rivers (see Box 6.6). Yet this happened in Papua New Guinea.

With this background about the power resources of business and the importance of the regime setting – and/or its North/South geopolitical context – it is possible to turn to an examination of the internal divisions within the business community and the implications that these have for environmental performance and responses to environmental conflict.

Although business can be expected to unite behind the core requirement that all firms need to continue to operate (including a common desire to limit the implications of environmental criticism and concern), business can be quite divided over the environmental implications of any given project or the economic consequences of any proposed piece of environmental regulation. For example, a firm pumping waste into a river will have a different attitude towards moves to limit water pollution from another firm downstream that uses the water in its processes. Such differences are reinforced and reflected in different attitudes towards the environment and environmental concern. For heuristic purposes, the attitudes of business can be divided into three groups (see Box 6.2).

These three categories of business – those who reject environmental concern; those who accommodate concern and change their practices or their marketing; and those whose businesses provide environmental

Box 6.2

The three attitudes of business towards environmental concern

1 Some sections of business comprehensively reject the case for environmental concern (rejectionists).

2 Others are largely sceptical but have sought accommodation by making limited changes (accommodationists).

3 Some have embraced the themes raised in environmental critiques and have redesigned their processes to minimise damage to the environment (environmental business).

Source: McEachern (1991: ch. 4)

choices – are not in themselves significant, as they merely describe the range of responses that business has taken to environmental concerns. They become important when considering the political response that business makes to politicised environmental concern. Here these positions and the relationship between them combine to shape and constrain the actions that governments are willing to take on environmental questions. The balance between the three is not fixed and is conditioned by the political complexion of government and the options presented on the policy menu. When governments hostile to environmental concern are in office, then the influence of the rejectionists is enhanced. When governments seeking to use environmental concern as an electoral resource are in office, green businesses have more scope for influence. Such shifts condition the response of the mainstream and change the substance that can be given to some very broad concepts seeking to link environmental concern with economic development.

The issue of climate change, amongst other issues and concepts, is now used to demonstrate how these different business responses are played out. Climate change is a global problem that transcends national boundaries and it is beyond the scope of any single nation or corporation to reverse the problem (Pearce 1991: chs 3, 4). The issue of international climate change exemplifies all the problems of scientific modelling of major system-wide changes. The problem can be simply described, at least as a hypothesis and then as a prediction of likely developments. Industrialisation has been based on the use of fossil fuels to produce energy. When burned, these fuels release carbon dioxide (CO_2) into the atmosphere. As energy consumption has increased, so have concentrations of CO_2 in the atmosphere. More atmospheric CO_2 increases the amount of long-wave radiation trapped by the Earth's atmosphere and enhances the natural greenhouse effect, which keeps the biosphere within a temperature range that is able to support life. On the back of evidence of increasing concentrations of carbon dioxide in the atmosphere, complex computer models of the global weather system and arguments about an enhanced greenhouse effect become predictions of global warming. Global warming can change the world's climate patterns and produce major changes in the world's environment. The world's mean temperature can rise, sea levels increase, deserts expand and climate bands shift, with major consequences for plants and animals.

It is important to understand some of the underlying complexity of the science and modelling involved in the prediction of global warming. Global temperatures fluctuated long before fossil fuel use increased, with periods of warming and cooling following 'natural' causes. Records of global warming or cooling are difficult to establish and are incomplete. Other factors complicate the position: natural processes like volcanic eruptions can increase the presence of 'greenhouse gases' in the atmosphere, but they also throw out particulate matter that has a cooling effect because of its ability to deflect part of the Sun's heat; increased air pollution and greater cloud cover (from the increased evaporation caused by higher temperatures) can also produce this cooling effect; and finally, the impact of the world oceans as heat reservoirs and CO_2 sinks is unclear. The hypothesis of the enhanced greenhouse effect is fairly simple, but the prediction of its consequences for global weather is based on two sets of complex computer modelling. The first deals with how the global climate systems work. The second models the processes that could link increased CO_2 concentrations to increased warming. In all such modelling exercises, the whole sequence of equations that make up the model are based on assumptions, hypotheses and odd bits of 'hard' information. Such uncertainty always makes the debate on the greenhouse effect complex and open to a variety of interpretations. Global warming is an environmental issue defined by its complexity, the ambiguity in the analysis and the long timescale for both effects and counter-measures.

This issue complexity allows business to interpret its response in different ways.

Rejecting environmental concerns

For the *rejectionists*, the case for environmental concern is, at best, a mischievous invention and more often a cloak to conceal the real intentions of those who are anti-business for a number of malign social purposes. Their goal is to protect business against this unjustified attack on the rights of private property and managerial prerogatives. To the extent that environmental damage is recognised, it is justified by the economic development that results and the importance of economic growth (both for improving standards of living and, if need be, to address serious instances of environmental damage). The business community has always had its share of colourful characters who are frank in their rejection of environmental concern. This comment by Hugh Morgan, the former CEO of Western Mining Company (one of Australia's major mining companies which was bought out by BHP Billiton in 2005) is typical of the rejectionist position:

> The road to power for ambitious revolutionaries is no longer the socialist road. But the environmentalist road, today, offers great opportunities for the ambitious, power-seeking revolutionary. Environmentalism offers perhaps even better opportunities for undermining private property than socialism.
>
> (quoted in McEachern 1991: 110)

When business is criticised for the harm it does to the environment, there are a number of propositions that rejectionists use to defend it. These themes form the bedrock of mainstream business efforts to limit the consequences of government regulation. The different themes are all arranged around a core claim about utility. Utilitarian arguments grew out of the forms of political economy dominant in Britain in the late eighteenth and early nineteenth centuries and are associated with the philosopher Jeremy Bentham. They hinge upon the notion that use and enhanced utility is the key to assessing value in an economic and social sense. In conflicts over the impact of business on the environment, it is the enhanced utility of production that justifies any harm done. The amount of damage is seen as far less significant than the enhanced utility of what business does. More goods are produced and circulated, standards of living rise, social wealth accumulates, and the ability of society to respond to environmental damage is enhanced. Here utility is linked to an account of the virtuous consequences of economic growth. It is important to note that this argument is usually constructed at the broadest level because it has a tendency to be unstable when confronted by the actual products produced and circulated. For example, the mining industry is willing to name products selectively and show their value. It will identify the role of coal in producing electricity and the good things that electricity does. It will also run through the range of useful products that are made from mined minerals, from battleships to aluminium foil. Nonetheless, justifying environmental damage is not so convincing when argued on the virtues of disposable polystyrene cups and packaging (recognised in practice by all who eat fast foods), disposable plastic shopping bags or the ever popular fluorescent shoe-laces. Trying to cite the utility of these products runs the risk of giving encouragement to a moral economy that is hostile to waste and the production of saleable but unnecessary products.

Linked to arguments about enhanced utility and the virtues of growth are claims about investment and employment. If environmental regulations or restrictions are imposed, then business can respond by claiming that both investment and employment will fall. Employment is used to justify whatever business is doing, since increased employment is a social good and unemployment is a social bad. Once again there is a dubious element in this claim, because much of what business does in its pursuit of increased profits also destroys jobs. For example, in forest disputes industry continually claims that logging restrictions will cost jobs. Environmentalists counter that the restructuring of the industry and the introduction of capital-intensive production technology has led to the closure of small saw mills and destroyed more jobs than environmental care. Here the rhetorical point of distinction is between the destruction of employment that limits profitability (environmental care) and the destruction of employment that enhances technical efficiency and boosts profitability. This invokes other arguments about economic efficiency and increased competitiveness, which are part of the everyday currency of contemporary political debates (see the discussion on technical and economic efficiency later in this chapter).

When arguments based on utility and the virtues of the economy are exhausted, it is always possible for business to turn to the claim that business does not harm the environment any more than is strictly necessary. More than that, business cares for the environment as much as, if not (certainly in a more practical and effective sense) more than, environmentalists do. Evidence for such environmental care may simply recycle claims about growth and the need to create the wealth necessary to fund responses to environmental problems. In other instances, the evidence for care can be based on the role of significant business figures in various conservation organisations and environmental campaigns. In Australia, for example, Sir Maurice Mowbray (a significant force in the mining industry) is noted for his role in promoting tree planting around mine sites as well as his involvement in the Australian Conservation Foundation. Similar examples can be found in the USA and Europe, especially now that various business organisations have set up specialist sub-groups to deal with environmental questions.

Wise and sequential use

The rejectionists' politics of business have become more sophisticated over time and a whole set of political initiatives have been created to enable 'business-as-usual'. Starting in the USA, and spreading by imitation, has been the development of grassroots anti-environmental organisations, frequently supported by business associations and firms and arguing under the rubric of 'wise use': an anti-environmental position. An account by Ron Arnold of the development of the US wise-use movement can be found on the Center for the Defense of Free Enterprise Website (url http://www.cdfe.org/wiseuse.html) This website also contains an account of the main beliefs of the wise-use movement.

1. Humans, like all organisms, must use natural resources to survive.
2. The earth and its life are tough and resilient, not fragile and delicate.
3. We only learn about the world through trial and error.
4. Our limitless imaginations can break through natural limits to make earthly goods and carrying capacity virtually infinite.
5. Human's reworking of the Earth is revolutionary, problematic and ultimately benevolent.

Perhaps a quote which best sums up the more messianic flavour of wise use philosophy is as follows:

> Humanity may ultimately prove to be a force of nature forwarding some cosmic teleology of which we are yet unaware. Or not. Humanity may be the universe awakening and becoming conscious of itself. Or not. Our reworking of the earth may be the utmost evolutionary benevolence and importance. Or not. We don't know. The only way to see the future is to be there.

There is an affinity between the wise-use perspective, other anti-government groups in the west of the United States (including the Freemen and parts of the gun lobby) and radical economic libertarians. Ron Arnold, for example, also served as Executive Director of the Center for the Defense of Free Enterprise (CDFE). This think-tank, according to Mark Dowie, an American critic of wise use, formulates 'an ideology for a grassroots insurrection he (Arnold) believes will save free enterprise capitalism from the scourge of environmentalism' (Dowie 1995a: 93). The work of wise-user Patrick Moore is also of interest here. Moore is North American and an ex-Greenpeace professional now working for industry. (For a clear statement of his position visit his website url http://www.greenspirit.com/ and read his informative biographical-political manifesto 'Environmentalism for the 21st Century'.) In 1998 Moore visited Australia, a guest of the Forest Protection Society which is an Australian wise-use-style group promoted by the National Association of Forest Industries (Strong 1998: 3). Moore was vociferous in advocating the prevention of the 'locking up' of forests in 'no go' zones in national parks and reserves, a characteristic wise-use position (Doyle 2000). In 2006, Moore again appeared on the Australian political scene, this time working for the US nuclear industry, promoting its 'carbon-free, clean fuel' credentials.

The significance of the wise-use movement does not lie in its philosophy but in the practical involvement of numbers of people in campaigns against expressions of environmental concern. In Australia, forest conflicts now frequently involve clashes between environmental activists, forest managers, forest companies and the Forest Protection Society, made up of such grassroots opponents of environmental concern (Box 6.3).

The wise-use movement has been assiduous in its advocacy of roundtable, consensus decision making. The Quincy Library approach has become a model for the management of environmental conflict. This approach had its origins when the town of Quincy, in the Sierra Nevada range, formed the group in 1993 ostensibly to resolve resource conflicts between timber and environmental interests, as well as government agencies (Sierra Club 1997: 1). Meetings were held in the town library so people would not yell at each other. The Quincy Library Group has been extremely successful in promoting its wise-use roundtable approach. Emerging from this process was The Quincy Library Group Forest Recovery and Economic Stability Act of 1997. It orders the US Secretary of Agriculture to hold a five-year pilot project based on Quincy style decision making on three national forests in the Sierra.

Box. 6.3

Wise and sequential use

1 Wise use shares many similarities to multiple use.

2 'No-take areas', exclusion zones, scientific control zones and ecological buffer zones are now increasingly obsolete. Under wise-use decision making, these areas are now referred to as resource 'lock-ups'.

3 Wise use is closely associated with the principle of sequential use. Any ecological 'use' is only considered after business interests have deemed it appropriate in so far as they have no further utilisation for the areas in question. The biosphere, in this view, can be used over and over again; fulfilling all the demands placed on it by the multitude of stakeholders, with no long-term negative consequences. Environmental interests and those of indigenous peoples are most often given access last.

4 Wise use advocates neo-liberal roundtable decision making to further its cause.

5 There is no labour stakeholder under wise use resource allocation. Under wise use workers and labour organisations are usually absent, now seen as a non-constituency, a non-stakeholder.

6 The role of the state is almost non-existent under wise-use. In many multiple-use models, the state is portrayed as a neutral arbiter between conflicting views and interests. In wise-use decision making, the state is, at best, usually just another participant in roundtable discussions.

7 Frequently, corporations initiate and control agendas and terms of reference for roundtable processes. They are largely self-monitoring and self-regulating.

8 Wise use attacks the notions of 'independent science' and 'ecology' and the organisations which traditionally provide it.

Rejecting climate change

A little girl blows away dandelion fluff as an announcer says, 'Carbon dioxide: they call it pollution; we call it life.' The television adverstisement continue with:

> The fuels that produce carbon dioxide have freed us from a world of back-breaking labour, lighting up our lives, allowing us to create and move the things we need, the people we love . . . Now some

> politicians want to label carbon dioxide a pollutant. Imagine if they succeed – what would our lives be like then?

The US Competitive Enterprise Institute ads aim to counter the media spotlight on threats posed by worldwide climate change (ENN 2006), and in particular the Al Gore documentary *An Inconvenient Truth* which was released a week after the the above advertisement premiered in May 2006. The advertisements directly counteract some of the arguments and images used in the 'alarmist' film.

Climate change has become the overarching environmental 'metanarrative', subsuming almost all environmental issues under its rubric because of its magnitude, urgency, and the fact that dealing with the problem does not necessarily have to interfere with free market economics. The latest studies have found the likely range of temperature increase this century will be between 2 – 11°C (Pittock 2005: x). Climate change will affect business directly

> via impacts on their activities, including raw material and production costs, insurance, prices and competitive position. They will also be affected by how greenhouse gas emission reduction measures may influence their costs and activities, and ultimately their competitive position. Whether or not business people believe that human induced climate change is happening, they will be affected by how their competitors, governments and society perceive the problem.
>
> (ibid.: 235)

Rejectionists have long held a position that climate change does not exist; that it was simply a scientific position dreamed up and then crafted by environmentalists to challenge business-as-usual, to restrict industry and development interests from making profits which were rightfully theirs. This is the 'climate denial' position. Industry-funded, right-wing think-tanks – such as the Heartlands Institute and the Institute for Free Enterprise – were particularly active in both Australia and the United States' eventual decision to withdraw support from the Kyoto Protocol process. In research aimed at monitoring and analysing these think-tanks' impact on the United States' climate change policy, McCright and Dunlap noted that: 'the conservative movement employs counterclaims to block any proposed action on global warming that challenges its interests' (McCright and Dunlap 2000: 518). Despite some of these claims being clearly nonsense – such as, the Kyoto Treaty 'would cut economic growth by 50 per cent by the year 2005' – they were backed by large public relations concerns with almost unlimited financial backing from multinational extractive industries. According to McCright and Dunlap, despite these arguments' lack of veracity, they 'played a decisive role in defeating the U.S. ratification of the Kyoto protocol' (McCright and Dunlap 2003: 367), proving that powerfully backed myths can beat scientifically 'correct' ones.

In Australia, the Lavoisier Group was established in 2000 to discredit climate change science and bring together business groups in opposition to limiting greenhouse pollution. They saw the negotiations leading to the Kyoto Protocol as an elaborate conspiracy in which hundreds of climate scientists twisted their results to support the 'climate change theory' in order to protect their research funding (Hamilton 2002). The Lavoisier submission to a Senate inquiry on ratification of Kyoto went so far as to compare it to the planned invasion of Australia by Japan: 'With the Kyoto Protocol we face the most serious challenge to our sovereignty since the Japanese Fleet entered the Coral Sea on May 3, 1942.' The Lavoisier Group continued to argue that environmentalists were the new 'imperialists' desiring 'to transfer, or destroy, wealth and income within Australia on a massive scale' (ibid.).

The group has been likened to the American 'Global Climate Coalition' which has struggled to retain its membership over the last few years as climate change evidence has become even more conclusive. It has lost major members such as Ford, BP and Shell (ACF *et al.* 2000), because many large businesses around the world now accept, at least, that they have to be seen as part of the solution to climate change.

On the flipside to the extreme antagonism to environmental concern of the Lavoisier Group and others, it seems likely that there are also a number of businesses who pessimistically believe it is too late to do anything to prevent disastrous global warming, and that we may as well persevere with 'business-as-usual' and wait for a technological fix.

Accommodating environmental concerns

The *accommodationist* position has become the mainstream position for business as the levels of environmental concern have increased and consumers have registered some level of environmental commitment in the marketplace. In a sense, some holders of this position try to maintain as much rejection as possible while accommodating key themes of environmental concern. Sometimes, the accommodation will be both strategic and rhetorical, as for example in the series of advertisements under the heading 'With Climate Change, What We Don't Know Can Hurt Us' published by Mobil to argue for a slow response to reducing greenhouse gas emissions (*The Australian* 6 Sept. 1996: 3). These ads do not reject environmental concern but seek to identify with it while arguing for a slow response, protecting the position of the energy companies and of energy-intensive industry. This position can also be seen in the various statements of the mainstream business organisations embracing sustainable development.

The politics of business's response to environmental concern and activist green organisations is not static. Different positions have developed to respond to the secular increase in environmental concern in society. Some mining companies, for example, which were in the past at the centre of rejectionist politics have responded in at least two ways. On the one hand, the mining industries in industrialised capitalist countries have adopted more integrated environmental management approaches in line with a commitment to sustainable development as outlined and identified by the World Business Council on Sustainable Development. There have been significant improvements in the ways in which mining companies in these countries have approached the task of better managing mining projects and the rehabilitation of mine sites. Excellent manuals are now produced to aid the work of environmental managers in the industry (for example, Mulligan 1996). This is a much more accommodationist stance than before. This also extends to the development of more open forms of environmental reporting as a standard part of business.

A lot of progress has been made in the development of a business perspective on environmental concern which is more than just a pragmatic adjustment to prevailing political and social fashions. Few have been as explicit in developing an alternative commitment as Paul Hawken who has campaigned tirelessly for business to take environmental concern seriously and to redesign all its business practices accordingly.

In his book, the *Ecology of Commerce* (Hawken 1993) he proposes an eight-point approach for business:

1. Reduce absolute consumption of energy and natural resources in the North by 80 per cent within the next half century.
2. Provide secure, stable and meaningful employment for people everywhere.
3. Be self-actuating as opposed to regulated or morally mandated.
4. Honour market principles.

5. Be more rewarding than our present way of life.
6. Exceed sustainability by restoring degraded habitats and ecosystems to their full biological capacity.
7. Rely on current income.
8. Be fun and engaging, and strive for an aesthetic outcome.

(pp. xiv–xv)

Undoubtedly this is a substantial challenge to business but Hawken uses his book to illustrate the practical way in which a variety of companies have sought to implement these principles. Examples range from reducing energy consumption, redesigning to eliminate waste and a commercial but environmentally effective response to toxic by-products. These ideas have been recast in terms of 'natural capitalism' and are gaining greater ground. This is a position associated with the work of Amory Lovins and Paul Hawken (Hawken *et al.* 1999) with an emphasis on re-engineering production. The aim is to eliminate waste by turning waste products into inputs for new processes of production, such that higher levels of efficiency generate both more profitable and more environmentally sound businesses. Such ideas are not universally applicable nor approved by all businesses or business consultants. Nonetheless, they represent a substantial move forward in the business position (see Box 6.4).

Sustainable development

The concept of sustainable development has had a relatively long evolution. In part, it has evolved in a manner which reflects business' move towards a more accommodationist position in relation to environmental issues. This position suggests that investment and growth can be reconciled with environmental concern. It became a central framework for discussing the interaction between business and the environment in the 1980s (after the publication of the Brundtland Report, *Our Common Future*, in 1987).[1] The emphasis on sustainable development was reinforced by the publication of *Agenda 21* and the wide publicity given to the Rio Earth Summit in 1992 (see Chapter 8).[2] Initially, business was suspicious of the concept, which it interpreted as being hostile to its primary concerns and giving too much emphasis to evidence of environmental damage and potential threats to the environment. Nonetheless, business organisations rapidly came to understand that adopting sustainable development was an effective response to environmental criticism, not in the sense that it changed how business conducted itself so as to do less harm, but because it provided a rhetoric to protect the continuation of business as usual. For example, by

Box 6.4

Environmental business awards

Paul Hawken, Preface to The Ecology of Commerce *(1993)*

On being awarded a Council on Economic Priorities 'Environmental Stewardship Award':

> I walked to the podium, looked out at the sea of pearls and black ties and fell mute. Instead of thanking everyone, I stood there in silence, suddenly realizing two things: first, that my company did not deserve the award, and second, that no one else did, either.
>
> What we had done was scratch the surface of the problem, taken a few risks, put a fair amount of money where our mouths were, but, in the end, the impact on the environment was only marginally different than if we had done nothing at all. The recycled toner cartridges, the sustainably harvested woods, the replanted trees, the soy-based inks, and the monetary gifts to nonprofits were all well and good, but basically we were in the junk mail business, selling products by catalogue. All the recycling in the world would not change the fact that doing business in the latter part of the twentieth century is an energy intensive endeavour that gulps down resources.
>
> I don't mean to decry the efforts made by companies to reduce their negative impact on the environment. I applaud them greatly. But it was clear to me in that moment that there was no way to get 'there' from here, that all companies were essentially proscribed from becoming ecologically sound, and that awards to institutions that had ventured to the environmental margins only underlined the fact that commerce and sustainability were antithetical by design, not by intention.

the time the Australian edition of *Our Common Future* was published in 1990, the Business Council of Australia, the Confederation of Australian Industries, the National Farmers' Federation and the National Association of Forest Industries had all committed themselves to sustainable development (WCED 1990: 37–53). Similar examples can be found for the USA and Europe. In the USA, the President's Council for Sustainable Development (PCSD) included business figures who made a similar commitment to sustainable development, although it should be noted that this happened only when the final report was released in March 1996 (see PCSD 1996: 1–10; Cushman 1996: 1).

The business commitment to sustainable development has not meant great policy innovation, but when business needs to present a case for environmental concern then it can be packaged using this concept. For example, Shell was praised for its executive policy statement on sustainable development in the early 1990s (Callenbach *et al.* 1993: 36–7) despite the concurrent impact of its actions on the Ogoni in Nigeria. In the Australian case, the enthusiasm shown by business for sustainable development as a defence of its existing practices, while evidence of continuing environmental harm was accumulating, discredited the concept. It was in this context that the Labor government invented the ESD processes, adding 'Ecological' to the notion of sustainable development as a way of symbolising renewed concern and a new way of getting environmentalists involved in the process (see Chapter 2).

One of the clearest and most advanced statements of the business vision of sustainable development comes in the publication *South African Environments into the 21st Century* (Hunter *et al.* 1989; see also Cock and Koch 1991). The book was produced with the involvement and support of the Anglo-American Corporation (the largest and most important company operating in South Africa). It was published as a response to the growing evidence of environmental damage accelerated by the economics of the apartheid regime. Anglo-American had been committed to debating the economic options facing South Africa and to exploring the requirements for a high-growth scenario as a response to the challenge of extending the benefits of economic development to the majority black population. Initially, it was concerned that a high-growth scenario could only be achieved with high environmental costs, and the book was an attempt to find ways of showing that only high growth could deliver both developmental and environmental benefits. The book consciously sets itself to address an elite of both white and black opinion, to reject a focus on wildlife (the frequent focus of white environmentalism in South Africa) and to concentrate on the human future and the impact of environmental degradation. It is a confident rejection of apartheid with an optimistic account of development, provided high growth is associated with an environmental ethic and government provides 'the discipline in the market' (Hunter *et al.* 1989: 122). This is as far as business thinking goes on the question of sustainable development. It is an important alternative to the frequently expressed belief in unfettered free markets, which has dominated in the West since the early 1980s.

Plate 11 ***Community-erected protest sign against nuclear waste dumping, Mitterteich, Germany. Courtesy of Chain Reaction***

Efficiency/profitability/technology as environmental care

Business, as individual firms, is able to respond to increased environmental care in a way that is both good for its profits and for the environment while not 'turning green'. Some kinds of economic and technical improvements can lessen the impact on the environment of crucial production processes. For example, it is possible for buildings to be retro-fitted to improve their energy efficiency and lower their running costs. This can boost profits, lessen the demand for electricity and lessen the production of greenhouse gas emissions from power stations. It is a little thing but no less important for that. In a related way, firms can adopt zero-waste strategies and concentrate on redesigning production so that waste products (including heat) can be turned into inputs for other processes, improving efficiency and lowering the impact on the environment. The integrated production system introduced by Xerox in the USA is an example. It is not that these changes substantially 'green' business, but they certainly lessen the impact of business on the environment while maintaining a business-as-usual stance, which is attractive to managers and shareholders.

The fact that some businesses make such changes shows that it is possible for business to show care for the environment, up to a point. The fact that it is possible does not mean that all business can or will make the appropriate changes. Nor does it mean that the same firm operating in a number of different countries will not pursue high-efficiency solutions in one setting and low-efficiency solutions in others; protecting the environment in one and harming it in another.

Accommodating climate change

The vast majority of businesses have now accepted the scientific and public consensus that climate change is happening, but have certainly not responded in a uniform manner. They range from those who have seized upon it as a green marketing tool (for example, the current 'remake' of the nuclear industry) to those who have had genuine roundtable discussions with environment organisations and the government to promote regulatory frameworks for reducing emissions and carbon trading systems. Many businesses have realised that they will have to change and adapt, and that early innovation will put them at a comparative advantage.

First, there are those who have 'co-opted' climate exclusively for profitable gain. Powerful business people who are still in climate denial now choose to advocate the position in more private circles. Their public utterances have now become more sophisticated, aided and abetted by right wing *wise-use* think-tanks and groups such as Frontiers of Freedom, the Clean Air Institute, Environmentalists for Nuclear Energy and the Clean and Safe Energy Coalition. These groups have changed tack and now help those in climate denial to 'green their products' for the purposes of 'improved positioning in the marketplace'. This stance can be called the 'climate co-option' position: we don't believe it . . . but, if we can't beat them, join them, and then beat them at their (the environmentalists') own game.

This 'climate co-option' repositioning has taken two main forms: first, by accepting climate change but remaining against the Kyoto Protocol as the 'right way' of addressing the problem; and, second, by advocating that nuclear energy is the answer to provide the Earth with greenhouse friendly fuel. Let us address the argument against Kyoto first. At the heart of this rhetoric lies the inaugural meeting of the Asia-Pacific Partnership on Clean Development and Climate (the AP6) in Australia in January 2006. As well as bringing together the two non-compliant nations – Australia and the US – to the climate change table, it also managed to secure the participation of China, India, Japan and Korea. The major theme of the conference was that it was with business, and not nation-states or the Kyoto Protocol, in which the salvation to a reduction in global climate change lay. This was continued in *The Weekend Australian* (14–15 Jan. 2006: 16) in an opinion piece without a byline, which reported that Kyoto meets the green lobby's 'ideological preference for bureaucratic solutions imposed on private enterprise'. Indeed this opinion piece continued with its own ideological rant:

> The reactionary response to the Asia-Pacific Partnership meeting this week demonstrates that support for Kyoto cloaks the green movement's real desire – to see capitalism stop succeeding. Extreme greens cannot bear to accept that our best chance of reducing greenhouse gas emissions will occur when free enterprise has incentives to implement solutions.
>
> (ibid.)

So, in this vein, 'extreme greens' are created who support Kyoto, and are portrayed as extreme opponents of free enterprise, whilst these think-tanks paint themselves as moderate greens, now accepting the current climate crisis but offering a different solution: a roundtable built by business; not the nation-state. Climate change will be resolved by the very perpetuators of the crisis out of some sense of corporate moral responsibility. In fact, this model is so attractive to big business industries, such as transnational extractive industries, that it will lead not quite to business-as-usual, but, some argue, to business better-as-usual. Cate Faehrmann of the New South Wales Nature Conservation Council argues that the APPCDC's six-nations voluntary approach was a 'license for government and business to do nothing . . . Without any incentives or penalities there is no reason for industry to move away from burning coal and fuel' (Faehmann quoted in Wikinews, 12 Jan. 2006).

The second 'climate co-option' argument is that nuclear energy is green, clean energy. The Nuclear Energy Institute (the main nuclear industry organisation in the US) has hired ex-Greenpeace professional and wise-user Patrick Moore to engage in 'grassroots advocacy'. Moore has been paid to set up the Clean and Safe Energy Coalition (Wald 2006), co-opting green and democratic language to further extractive industry's gains. In a statement to the US Congressional Committee in 2005, Moore made the climate/nuclear link the primary argument of his presentation, one which continues to inform his wise-use style, public relations work for the industry: 'A significant reduction in greenhouse gas emissions seems unlikely given our continued heavy reliance on fossil fuel consumption. An investment in nuclear energy would go a long way to reducing this reliance' (Moore 2005).

Again, in Australia, Moore has been active greenwashing big business interests. He was selected by *Sixty Minutes* reporter Peter Overton to provide the 'balance' to his story on 'The Nuclear Solution'. Moore was identified for the purposes of the story by using his ex-Greenpeace credentials, but his affiliation to the powerful nuclear-industry-funded lobby group was not (Overton 30 April 2006). Again, the climate change argument was the central premise to Overton's story, with nuclear energy posed as the key potential solution. Climate change is now so feared by the public that it has enabled the nuclear industry to paint itself as 'carbon-friendly'.

However, not all businesses have co-opted green marketing techniques in order to benefit from the looming climate disaster. Unlike the AP6 which only focused on voluntary, non-binding measures to secure future greenhouse gas reductions in the region through accelerated development and technology transfer (EcoGeneration 2006: 7), many businesses actually called for ratification of Kyoto and demanded other regulatory market-based measures to encourage private sector investment and innovation to reduce greenhouse emissions.

One such business roundtable in Australia, comprising BP, IAG, Origin, Swiss Re, VISY and Westpac, together with the Australian Conservation Foundation, demand just such mechanisms as they agree that 'Climate change is a major business risk and we need to act now' (Australian Business Roundtable on Climate Change 2006: 1). They suggest that business and governments work together to design a 'long, loud and legal' framework such as introducing a national market-based carbon pricing mechanism for cost-effective emission reductions, setting a short-term binding target for Australia in 2020 to 'facilitate a smooth transition to a low-carbon economy' and as a milestone towards achieving a large, long-term, emissions-reduction goal, encouraging innovation and investment in emerging and breakthrough technologies (ibid.: 7).

Carbon trading refers to 'the trade in "rights to pollute"', be they in the form of pollution quotas set by governments or 'credits' generated from offset projects' (Ma'anit 2006: 4). Pressured by industry, the state governments of all the Australian states have already established carbon trading schemes. These follow in the wake of other international business initiatives such as the Chicago Climate Exchange started by Ford, DuPont and BP America to 'administer a multinational and multi-sector marketplace for reducing and trading greenhouse gas emissions' (Pittock 2005: 239–240), and the EU Emissions Trading Scheme.

These schemes can be viewed as commendable responses from typically conservative businesses who are finally fulfilling their environmental responsibility charters, as they will cut greenhouse emissions while not upsetting the status quo of business and market operations. Alternatively they can be seen as an unwanted distraction from actually taking responsibility for climate change and actively preventing it (Ma'anit 2006). This climate accommodation and co-option by business may not amount to much if worldwide consumption continues to grow unabated, promoted by these very same companies and neo-liberal economic systems.

Environmental business

The *environmental business* position is increasingly common, but is frequently under attack from mainstream and rejectionist positions alike. Environmental businesses are thought to have conceded too much to the environmental critics and to undermine the positions taken by the rest of the business community. It is now possible to find a range of companies from those, at the very least, which have used environmental redesign as a way of marketing their products, to those which have redesigned their production processes in attempts to reverse the state of environmental destruction. Companies along this spectrum include businesses such as Ecocern which provides 100 per cent post-consumer waste recycled paper, to companies involved in pollution clean-ups or environmentally friendly energy companies. Even firms within the same industry will take different 'green' paths (see Box 6.5 and 6.6).

Environmental business also includes attempts by the corporate sector to alter the 'mind-set of business leaders towards the environment', as is the case in aspects of business management education in North America.

Box 6.5

Green wine or green marketing? The different faces of accommodation

Similar to the shake-up of the agricultural industry, environmentally aware consumers are now seeking out sustainable viticulture and 'clean green' wine. However, while only a very few wine businesses have gone completely organic, most are now trying to project an environmentally positive image and accommodate green concerns.

The distinction can be seen quite clearly in the contrasting South Australian case of BRL Hardy Wine Company's Banrock Station property and label of the same name, and the Temple Bruer organic winery further south.

Temple Bruer became certified organic in 2001 by the Biological Farmers of Australia. The owners strongly believe that organic farming 'should be the "standard" way of farming here in Australia' (Hankin 2002:13), and they have modified all their practices accordingly, including using no insecticides for 24 years (Bruer 1999: 12). Beyond that, they recycle all their grape waste into an on-site composting system, use drippers on all vines and water monitoring systems, process waste water through a solar-powered still and recycle it, and have created mass replanting programmes including replanting 30 per cent of the vineyard and a 500 m long roadside verge with native vegetation and trees. They hope to ultimately 'sequester', in trees grown on the property, all the CO_2 generated by the winery so that Temple Bruer offsets all its greenhouse gas emissions (ibid.).

Banrock Station, on the other hand, claims cutting-edge advances in environmentally friendly viticulture and business, but is such a large industrialised winery that its 'greenness' comes more from giving large amounts of money to environmental projects than in reforming its practices. It claims to manage the vineyards using sustainable horticultural practices (Dowling 1998: 23), but still uses chemicals as going organic would evidently affect its bottom line. However, like Temple Bruer, it does use drip irrigation and water monitoring, grows cover crops between vine rows, has established woodlots watered by the winery's effluent, and has removed livestock and regenerated native vegetation. Yet it tends towards larger 'showier' environmental projects, such as building a $1 million visitor centre using environmental building technologies, working in partnership with Wetland Care Australia through the Banrock Station Wetland Liaision Group, and donating part proceeds of every cask to Landcare Australia for wetland rehabilitation (ibid.: 22–23).

This is a good model of the visible ways in which accommodation with environmental concern can be pursued.

Box 6.6

The international fight for Jabiluka and shareholder activism

From the late 1990s, the mining company Energy Resources Australia (ERA) and its major financiers of North Ltd and Westpac bank attempted to open Jabiluka uranium mine in Kakadu national park, northern Australia. The successful campaign against the mine was able to mobilise a huge network because of its tripartite argumentation: it was firmly meshed with the land rights concerns of the indigenous Mirrar people, Kakadu national park has huge wilderness conservation value and the fact that it was a uranium mine drew in the anti-nuclear movement (Doyle 2005: 77–8).

Along with an eight-month blockade at the mine site in 1998 involving over two thousand people acting under the leadership of the traditional owners through the Gundjehmi Aboriginal Corporation, Jabiluka Action Groups in many of Australia's urban centres used various boycotts and direct action against the corporations involved. Opponents of Jabiluka bought shares in North Ltd and formed a 'North Ethical Shareholders' bloc which used the October 1998 shareholder meeting to insist the company hold an Extraordinary General Meeting focused on Jabiluka (Sykes 1999). In order to prevent North from raising the money required to fund its milling operation at Jabiluka, activists also targeted North's major investor Westpac through protests at local branches and the 1999 AGM. North then pulled out of the project and Rio Tinto became the majority shareholder of the mine, but dissent from traditional owners, ongoing bad publicity and falling uranium prices led them to eventually abandon the project in 2003.

Businesses are increasingly having to deal with environmentalists and the impact that activism has on their image, rather than dealing directly with the state as was frequently the case in the past.

Source: Adapted from Doyle, T. (2005) *Environmental Movements in Majority and Minority Worlds: A Global Perspective*, Rutgers University Press, New Brunswick, New Jersey and London; and Sykes, T. (1999) 'How Activists Have Hijacked the Annual Meeting', *Australian Financial Review*, 11 July

Greening business management education in North America?

The long history of the Environmental Protection Agency in the USA, and the pressure exerted by numerous environmental campaigns, has made US business particularly sensitive to the challenge of politicised environmental concern. In this context, US business has developed a

Plate 12 ***Arundhati Roy addresses tens of thousands of participants at the opening night plenary of the 2004 World Social Forum, Mumbai, India. According to some estimates, by its final day, the Mumbai World Social Forum had 140,000 attendees. Courtesy of Joel Catchlove***

number of accommodationist strategies to contain environmental criticisms through the use of political organisations and lobbying muscle. This section considers the reactions of US businesses that have responded positively to environmental concern by reviewing their production processes and seeking technical efficiencies that deliver both better environmental outcomes and improved economic viability.

The best illustrations of these gains come from a consideration of the recognition and treatment of waste. It is a commonplace to note that waste and pollution are basically resources that are not being used. So for example, the smelting of sulphur-rich ores releases vast quantities of sulphur dioxide and other sulphurous compounds. It has been common for these by-products to be released into the atmosphere as a waste, contributing to acid rain and damaging vegetation and soils. Under certain circumstances, smelting firms recognised that these sulphur compounds could be used as a raw material for other production processes (for example, they could be converted to sulphuric acid and sold as a raw material to fertiliser manufacturers). Capturing and selling these compounds would therefore increase the profit from each tonne of ore, as well as reducing the amount of environmental damage. In this sense, there is a more economic and ecologically efficient use of a non-renewable resource. A further example would be the capturing of heat released by one process so that it can be utilised by another process or firm. This synergy (one industry's waste product becoming the raw materials for another) can be generalised as a model for business responses to the environmental costs of its activities. The commercialisation and interconnection of by-product flows between firms can be proliferated seemingly endlessly. Such arrangements have been given the title 'industrial ecology' (Lowe 1996: 437–71).

An interesting question arises at this point. If such efficiencies and opportunities for increased profitability exist, why have they not been recognised until now? More importantly, what factors or forces can produce the changes necessary to make business see its waste either as a sign of inefficiency or as a potential source of revenue? One possibility is in the 'greening' of managerial education, so that decision makers will recognise such opportunities and act on them. The role of such education is very important in the USA, where there is a well-established tradition of MBA training for managers. This system is highly significant for three main reasons: first, many of the world's major international corporations base their head offices in the USA; second, many transnational firms recruit executives from the USA; and third, the American style of management training is being imitated in many other industrialised countries (Australia included). It appears that significant elements of environmental management have been introduced into the US managerial training system. For example, environmental case studies are being researched and taught at the Harvard Business School, the Haas Business School (University of California at Berkeley) and the Management Institute for Environment and Business (Washington, DC). A number of books have also been published worldwide to extol the important business opportunities that lie in better environmental practices. These typically set out examples of good commercial responses to questions of policy design and waste management, promoting integrated production and environmental accounting (see for example Callenbach *et al.* 1996; Ditz *et al.* 1995; Fischer and Schot 1993; Hawken 1993; Lowe 1996). The International Standards Organisation is also developing specifications for monitoring and measuring the environmental performance of business. In Australia, the Commonwealth Department of the Environment has compiled an on-line *National Cleaner Production Database*, which cites examples of firms that reduced both costs and pollution.

It is not just managerial education that produces such a change in approach. Some modes of government regulation, in conjunction with the pursuit of competitive advantage, have also impelled closer attention to environmental standards and actions. For example, car producers in Japan, forced to adopt tougher emission efficiency standards, have found that this helps them sell in the European and US markets as environmental concern and regulations tighten in those countries. Similarly, US chemical companies working on substitutes for CFCs were able to use negotiations over ozone regulation to give them a competitive advantage over international rivals. Even environmental impact assessment processes, entered into for public relations purposes, can play their part in making business sensitive to the advantages of changed ways of producing and using waste products, internalising forms of ecological rationality in the process (see Chapter 7).

Mitigating climate change

Many companies now directly profit from being early leaders in promoting low-carbon technological solutions to climate change (solutions that meet consumer demand rather than decrease consumption, which businesses often tend to argue is the government's responsibility). These include solar photovoltaic cell manufacturers, wind turbine manufacturers, electricity companies that deliver the option of renewable GreenPower and many others in the electricity sector.

With the advent of 'peak oil' (the instance when the worldwide extraction rate of oil reaches its highest level and thereafter begins its decline, which many experts believe will occur in the near future) (Post-Carbon Institute 2006), some oil companies are now researching and developing alternative biofuels. One such company is South Australian Farmers' Fuel (SAFF), who were the first government-approved and operating biofuel retailer in Australia. Their products use bioethanol from wheat and sugarcane, which is purified and dehydrated alcohol is used in petrol fuels (SAFF 2006). Similar to the industry progression from leaded to unleaded fuel, these biopetrols produce 10 per cent less greenhouse gas emissions than regular unleaded fuel. While the heavily green marketing is perhaps not always justified, SAFF are also currently researching ways to turn many other waste products (for example, human sewage) into fuel (SAFF rep. 2006), which could have other environmental benefits within the current system. There are also some negatives associated with biofuels as their production competes with the food-producing capabilities of limited arable lands, and tends to support the industrial development of vast agricultural monocultures.

Plate 13 ***Food sovereignty banner at the 2004 World Social Forum, Mumbai, India. Courtesy of Joel Catchlove***

Conclusion

Business power is one of the significant factors affecting both the degree of harm done to and the degree of care for the environment. With business pushing for a competitive edge and profitability, it can have strong incentives to reduce the amount of time and money spent on minimising the impact of economic development on the environment. However, if business can be persuaded of the need to factor environmental costs into its calculations, it can be a potent force for limiting or repairing ecological damage. Here, the pursuit of commercial advantage can sometimes drive business towards greener options and outcomes. The important question remains: under what conditions, in what circumstances and to what extent will business seek to harmonise its profit-seeking objective with protecting and improving the environmental condition of the world?

Further reading

Blair, A. and Hitchcock, D. (2001) *Environment and Business*, Routledge, London and New York.

Bansal, P. and Howard, E. (eds) (1997) *Business and the Natural Environment*, Butterworth-Heinemann, Oxford.

Frankel, C. (1998) *In Earth's Company: Business, Environment and the Challenge of Sustainabiliity*, New Society Publishers, Gabriola Is., Canada.

McEachern, D. (1991) *Business Mates: The Power and Politics of the Hawke Era*, Prentice-Hall, Sydney.

Institutional politics and policy making: the greening of administration

- **Institutions, bureaucracy and the legal frameworks**
- **Ecological and administrative rationality, and governmentality**
- **Public sector reform and market-like instruments**
- **Local administrative responses to environment: public inquiries, citizens' juries, and local *Agenda 21***
- **International environmental institutions, conventions and regimes**

Introduction

No matter how much environmental politics has focused on pressure group/NGO activity, the development of green parties and the use of environmental issues as part of the electoral contest between rival political parties or the activities of business, much of what happens in protecting or harming the environment comes from the management and making of policy in the hands of administration and bureaucracy. Whatever governments do, whatever laws or regulatory measures are introduced, these have their impact on the environment through the actions of various government departments and agencies. When considering the impact that politics has on the environment, much attention needs to be paid to the structure of administration, especially the interaction between departments and agencies with different responsibilities. Public administration is another site where the dispute between the demands of economic development and environmental concern is played out. The institutional design of the system of administration – whether local, state, national or international – is important for what is done, since it can support or inhibit moves to take greater care of the environmental consequences of economic development. Some versions of institutional design may be more effective for having environmental concern written into the detail of a policy and into the practice of its administration. It is not just a question of the relationship between different government agencies but also a question of modes of rationality and the 'administrative mind'.

This chapter begins with a consideration of the nature of bureaucracy, the kinds of policy assumptions that are embedded in its normal modes of operation and the limitations these impose on environmental initiatives that come within its grasp. It then goes on to consider the 'battle' for the administrative mind and reviews the different instruments used to make and shape environmental policy, including an account of the currently fashionable use of market-like instruments. Discussion then moves to institutional politics and policy making. These institutional responses range from every level of governance: local, state, national and international. At the local level we investigate how some local-level bureaucracies have responded to environmental concern, as well as reviewing citizens' groups which use administrative tools to further their environmental case. At the other end of the bureaucratic continuum are international institutional and administrative responses to the environment. These include diplomatic conventions, conferences and legalistic frameworks, creating international and transnational administrative agreements between nation-states, relating to ozone depletion and climate change, population control and hazardous waste management.

The politics of bureaucracy

What is bureaucracy and what is its place in the political and policy-making order? This is an important question, and confusion often surrounds the answers. The formal descriptions of constitutions and popular political explanations can often be at variance with what happens in practice. Bureaucracy is the name given to those parts of the state that are principally charged with the administration of government decisions and programmes. This implies a simple and clear model of the state system and the interaction of its various parts. Government is at the centre of the system making the political decisions that define policy initiatives and the form in which they will be achieved, raising revenue and distributing resources through its various programmes and agencies. It is the responsibility of the bureaucracy to administer the decisions made by government. (The legal system, the police and the military also have their parts to play in the enforcement of government decisions and the rules of the system as a whole.) It is implicit in this description, and is occasionally stated as such, that the bureaucracy is not responsible for the ends of its activity, it merely neutrally administers the decisions of others as efficiently as possible within the resources made available by government. Such a view of a neutral administration can be misleading. It is certainly true that politicians make the key decisions, but the bureaucracy also has a part to play in providing information, advice and assessments of different policy options. As the 'Yes, Minister' interpretation in the United Kingdom suggests, such advice can be quite significant in shaping, if not the decision itself, then the form of the decision and many of its finer details. Further, the bureaucracy might well have interests and concerns of its own that may not be the same as those of the politicians who are in charge. The tension between such divergent interests can have its impact on the making and administration of environmental policy.

This image of the decision/administration relationship is most appropriate to countries like the USA, Australia, Canada, Britain and other European countries. It is not clear how well this relationship model applies to countries with authoritarian and military regimes. Certainly, in countries where the pay of public servants is both low and irregular, it is unlikely that administration has the capacity to carry the decisions made by government. Other inefficiencies can appear due to different degrees of corruption, where enforcement becomes less systematic and impartial.

The foremost theorist of public administration and the politics of bureaucracy is Max Weber (1978). Weber is important for his foundational analysis of the 'rationality' of administration and its ability to produce an efficient linking between intention, means and ends. He is also important because later in his life, as he wrestled with the consequences of what he saw in administrative practice, he became an angry, depressed critic of what he described as 'the iron cage of rationality', the imperialistic application of administrative rationality to more and more aspects of life.

As far as Weber was concerned, administration achieved both its efficiency and danger from its ability to impose rationality on the performance of various tasks such that the personality, interests, associations and sympathies of the administrator would be stripped away and made irrelevant to the task at hand. The performance of a task could be codified in a set of rules and procedures, minimising discretion and maximising the certainty of performance.

Ecological rationality

The key assumption being made in Weber's analysis concerns a uni-directional conception of rationality. Although rationality has no content, its application renders predictable decisions that would otherwise be random or partial. 'Administrative rationality' of this kind (linking means to ends and dividing tasks into smaller, interlocking components so as to make decisions into rules and procedural boundaries) carries whole sets of assumptions about how to define the ends of administrative action, what counts as evidence, and how to measure technical efficiency in decision making.

If it is assumed that there is 'administrative rationality' of this kind, how does it relate to the 'rationality' of economic or marketplace decision making and how does it relate to the administration of environmental care? The relationship between 'administrative rationality' and the prerequisites of capital accumulation and economic growth is ambiguous. On the one hand, it is possible to find a positive relationship based on the conservatism associated with the practice of public administration. The social values implicit in the life-world of bureaucracy may embrace a given economic and political *status quo* and effectively serve the defined ends of that consensus. Hence, if that *status quo* is based on either a respect for private property (as in the USA) or respect for state ownership (as in the former USSR), in a context that favours economic growth over environmental concern, then bureaucratic conservatism will embody that ordering of priorities. In such circumstances, environmental concern, to the extent that it is given an administrative dimension, is likely to be handled in a way that favours this prevailing set of social/political assumptions. Radical and challenging environmental initiatives will be blunted and contained by the embrace of administrative rationality.

On the other hand, the logic of public administration does not rest on the same set of assumptions as that of entrepreneurial, competitive economic activity. Attitudes to risk and innovation can be quite different, and the whole spirit of pursuing ordered administration, detail and documentation could well act as a brake on the rationality of the marketplace. There could be antagonism between administrative and economic rationality. If that were the case, then there would be a basis for the effective administration of environmental concern. Indeed the same kind of administrative conservatism could provide an opening to environmental concern through extended/renewed regulation and supervision of economic conduct.

In exploring this puzzle, John Dryzek (1987, 1990, 2005) has posed a contrast between administrative and ecological rationality. Accepting Weber's account of administration as a way of conceiving rationality, Dryzek sets out a series of assumptions that take into account the principles of sustainability and what kind of ecological calculations these require.[1] On this basis, he then asks what it would mean for ecological rationality either to supplant or to supplement prevailing patterns of economic or administrative rationality.

At the heart of Dryzek's project are strands of optimism and pessimism. Pessimistically, Dryzek shows how strongly entrenched are both administrative rationality and its attendant forms of cost–benefit calculation. Optimistically, he shows how little needs to be changed before the organisational logic of bureaucracy starts to grind out better forms of policy making once ecological rationality gets into the process.

It is possible to make this argument because the way in which administration proceeds is compatible with the pursuit of newly defined social goals. Once the political process has placed environmental concern on the agenda and has enacted laws and regulations, the bureaucracy (which is neutral about ends) will render these administratively efficient. More than that, the way in which the bureaucracy supervises and regulates social conduct is compatible with the task of environmental regulation. Indeed the calculations, the range of information that needs to be gathered, assessed and applied, are but an extension of the normal mode of bureaucratic activity.

Establishing the balance between these two possibilities brings us to the debate over environmental intervention and the administrative mind.

The administrative mind

In a major work edited by Paehlke and Torgerson (2005), the consequences of bureaucracy are probed for their impact on achieving environmental goals. Here the battle is between co-option and transformation. On one side is the view that once environmental politics gets into the administrative embrace any far-reaching, radical, social and political implications are lost. More than that, the logic of administration means that environmental objections and environmental concern are broken down into administrative pieces and harmonised with the prevailing goals and assumptions of economic growth. Hence, there are accounts that show how environmental impact assessment (EIA) processes became devices for saying 'yes' to development with, at most, incremental adjustments to make environmental concern serve the ends of development. Hence, expressions of environmental concern are used to fine-tune and justify development plans, not to replace them with environmental care itself.

Against this interpretation, there is a more optimistic view that emphasises the corrosive effects of ecological rationality upon the ordered decision making of the administrative state. Instead of administration colonising the environmental project, the environmental project comes to colonise the administrative mind, displacing economic rationality by ecological rationality, which is then driven deeper into social and political processes by the routines of the administrative mind. Once again, EIA processes can be used to show how all this works.

Companies, to comply with environmental regulations, agree to undertake EIAs for purely pragmatic, instrumental reasons. Indeed, they may agree with the most cynical and shallow of intentions. Nonetheless, in the process of producing EIAs, and with the expectation that further EIAs will be required, the internal organisation of a company and of its calculations is altered. Part of the company may be given the task of producing or monitoring environmental performance. The public relations section may be given the task of overseeing the presentation of the company's case to a sceptical world. Some member of the board or senior management may be assigned responsibility for a newly created environmental portfolio. It is possible that within these slight, even cosmetic, changes the seeds of an ecological rationality can be sown. For example, a company may, as a result of being forced to produce an environmental audit of its internal processes, come to recognise that it has been wasting valuable resources as pollution. As a result, a change in the production process could be ordered to increase economic and technical efficiency, which has the consequence of reducing a harmful impact on the environment (for a more detailed account of the economic benefits of EIA, see Thomas 1996).

It is always hard to trace such changes to a prime cause, for example the use of EIA procedures to justify development. It is slightly easier to show that such changes have been taking place in some companies in some countries at some times. It is not that ecological rationality has replaced administrative, economic and other forms of prevailing calculations, but even the addition of some elements of ecological rationality lessens the harmful impact that development can have on some aspects of the environment.

Governmentality

There is one body of recent theory that focuses precisely on the processes by which 'new' mentalities come into being and shape the interaction of government, companies and society. That is the literature on governmentality. 'Governmentality' was a concept proposed by Michel Foucault as a way of understanding the form of government that occurs in the contemporary era when coercion and regulation are not the only means used to produce ordered, predictable, manageable behaviour (Foucault 1991). This very basic idea has been taken up by others and used to understand the 'mentality' that makes people or an issue 'subject' to government. Miller and Rose (1993), for example, trace the way in which a proposed new accounting standard spreads and produces new modes of calculating economic performance, which then act to promote investment and growth. The introduction of these reforms does not rely on the direct regulation of economic activity by government but still produces regulation-like effects. Just as new accounting procedures can transform calculations, so can new ways of taking environmental considerations into account. An environmental problem, pollution for example, can be recognised. New modes of calculation and measurement can be introduced to produce the information needed for a solution to be found. A 'solution' then emerges from diverse places in the social world, not just from government. As a result, the 'problem' has been made amenable to decentralised forms of regulation.

One of the central themes of Foucault's initial lecture on governmentality concerns the place of the state in theorising both government and policy.[2] Foucault's purpose in seeking to theorise governmentality was to call into question the central role ascribed to the state in the understanding of government in the contemporary era. Foucault wanted to move away from a conception that saw sovereignty, government and policy making flowing down into society from the central command point of the state. Instead, Foucault wanted to suggest that the state was but one site for the promulgation and policing of policy and the conditions that governed the conduct of populations. His is an account of society treated as possessing numerous diffuse sites for initiative and resistance.

In the field of environmental conduct, there is a sense in which we can trace through this diffuse pattern of policy making and policing the conduct of the population. Consider the situation in both the first Rio Earth Summit in 1992, and the subsequent Earth Summit 'Plus Ten' in 2002, when the media was saturated with images and programmes of enhanced environmental concern, frequently focusing on the behaviour of individuals either as the cause of environmental problems or as the key to solving them. Business, advertisements, green groups, media outlets, politicians and other social forces all combined to send out messages about the need for individuals to change the way they behaved and to internalise and police themselves on the observance of these new rules for environmental conduct. This was not a question of a state edict or regulation but of diffuse sites of social power and influence propagating various messages that, as they intersected, mapped out new patterns for social conduct. This focus on the character of persons, and the factors that govern their routinised and normal conduct, complements and intersects with other social projects, changing the form of ecological calculations used in debating and assessing the environmental conduct of business.

Although there are problems with the governmentality literature (including an overemphasis on functionality, order and regularity as well as a failure to identify and analyse the mechanisms that produce new or unstable patterns of governmentality), the basic insight adds to our understanding of what has happened as environmental concern and regulation have become the subject of more extended debate. It certainly produces a different way of understanding the consequences of administration in the making and maintaining of policy.

Economic and market-like instruments

The possibility that environmental concern could reinforce the urge for bureaucratic action has been noted by free-market economists, especially those associated with various right-wing, pro-business think-tanks. Their response has been to adapt their general critique of state intervention to this situation and to promote alternative, more market-like, approaches. To support their rejection of environmental regulation, there has also been a tendency to reject environmental concern itself as either fraudulent, misguided or based on far too complex and uncertain science to justify determined government action. It is instructive to revisit this critique of environmental regulation (crudely portrayed as a system of command and control) and the alternatives of 'flexible', 'market' mechanisms.

The use of direct government regulation to influence the impact of business (say) on the environment has been a common response to both particular and general environmental problems. For example, the decision to ban DDT because of its propensity to accumulate in the food chain is an example of 'direct' regulation. The government issues an order, a regulation, prohibiting the use of a particular substance except in prescribed, regulated or licensed circumstances. Such is the case with the use of ozone-depleting substances banned under an international protocol. For many years, efforts to reduce air pollution were based on state regulation of the amount of particular substances that could be released into the environment, with whole regimes of inspection and prosecution built around that regulation.

Despite the ease of promulgating regulations, there are problems with direct regulation, especially in the areas of policing compliance and in the unintended consequences that can follow from rigid enforcement. Often, regulation proves ineffective when enforcement is lax (which can be caused by a whole range of factors ranging from the regulatory agency being underfunded to political pressure to ignore regulatory breaches). Sometimes, it is claimed that the cost of regulation is not matched by the benefits that flow from it. Sometimes, the cost of compliance is high and firms may act to evade such regulations. Sometimes, the chances of being detected and punished for a breach are so low that there is little incentive to comply (although the US is renowned for its high rate of environmental litigation challenges – see Box 7.1). All these factors undermine the efficiency of regulation.

The questions of rigidity and unexpected consequences are much more difficult and serious. An oft-cited example concerns a US decision to impose the fitting of 'scrubbing' devices to coal-fired power station chimneys in a move designed to limit air pollution in general and sulphur emissions in particular (Cairncross 1991: ch. 5). Although such a regulation falls equally on all coal-fired power stations, its impacts are costly and counter-productive because it handicaps high-polluting and low-polluting power stations alike. This forces up the price of power when, other things being equal, low-polluting power stations using low-sulphur coal could have responded more flexibly and more cheaply to the challenge. Such examples can be multiplied to show that bureaucratic regulation can be expensive and unable to deliver acceptable levels of environmental outcomes.

This account of bureaucratic inefficiency, tendentious though it may be, fuelled the thought of those wedded to neo-classical economics and inspired a whole range of measures to deliver the same kinds of environmental outcomes with greater flexibility and lower costs.[3]

Box 7.1

US law and the environment

There is no more litigious environmental movement on the planet than found in the US. Since the 1970s' movement professionals have placed a heavy emphasis on environmental action pursued through the legal system. In the 1970s environmental groups had many key laws passed and amended, including convincing government agencies of the need to write environmental impact statements (McSpadden 2002: 175). Between 1990 and 1996, Justice Department statistics list 262 separate environmental lawsuits being filed. Even more dramatic are the figures since June 1997, with 184 cases filed (Paige 1998: 16). This reliance on the legal system is partly based on the fact that the US legal system is sophisticated enough to handle often complex, ecological disputes. Less positively, this cacophony of litigation has seen the development of a plethora of one-off victories. As such, the United States is not a society where we have seen the emergence of a permanent parliamentary or administrative environmental response. For example, as aforesaid, the success of green parties in the US is virtually non-existent, and there is no long-standing cabinet-level portfolio in national government. This reliance on the legal system is partly responsible for the lack of development of a really lasting administrative infrastructure built to deal with environmental issues on a daily basis.

However, the US legal system is a useful tool for environmentalists, especially in its broad interpretation of 'standing' which gives non-human beings much greater consideration in the eyes of the law. Famously, the rights of the old-growth-forest-dependent northern spotted owl have been challenged in various legal and political battles. The owl was granted protection under the Endangered Species Act in 1986 after habitat loss through excessive logging, but it was only after environment groups sued the government that President Clinton issued the 1994 Northwest Forest Plan which restricted logging in a 2000-acre radius surrounding all known spotted owl nests (Dawdy 2000). The timber industry claimed this cost 85,000 jobs (Fitch 2002: 432). Yet the spotted owl did not recover, and in 2004 another court trial came to an end when the 9th Circuit Court of Appeals overturned US Fish and Wildlife Service's interpretation of the Endangered Species Act. The court ruled that the Endangered Species Act was 'written not merely to forestall the extinction of species . . . but to allow a species to recover to the point where it may be delisted' (Eureka Times-Standard 2004). Now a similar case has been launched by environment groups to protect the spotted owl in Canada (Sierra Defense Legal Fund 2005).

The most common example used to illustrate this approach concerns strategies for pollution abatement. Instead of simply banning or limiting the emission of particular substances, a market for tradeable pollution permits can be created. First, some estimate is made of the total quantity of pollution being produced in an area or by a particular industry and a target is set for a lower level. Second, a number of permits are created that entitle each holder to emit pollutants up to a certain level (these entitlements add up to the total target level of pollution). Finally, these permits are then either issued, sold or auctioned off to polluters. Firms can then choose a flexible response to the problem of their polluting activities. They can simply buy enough permits to cover their current level of emissions because they find it cheaper than investing in new plant. They can buy fewer permits and reduce their emissions to the purchased level by investing in end-of-pipe pollution control devices. They can buy the permits they need now, invest in less polluting production technology later and resell surplus permits to other firms when their levels of pollution decline. The advantage of this scheme is that both the overall level of pollution is reduced and the price of the permits gives firms an economic incentive to reduce their production costs by reducing pollution. Some firms will find it cheaper to invest in pollution reductions, others will have to pay for more permits. In the end, all firms are required to pay for part of the costs of their environmental damage by buying permits. Cleaner firms pay less and dirty firms pay more (this at least partly internalises a previously neglected negative production externality). Another feature of this scheme is that it is possible to gradually reduce the aggregate pollution level further by either withdrawing permits, reducing each permit's pollution entitlement, or buying back permits to take them out of circulation. In this way, the scheme can further harness market forces and the price mechanism to the task of reducing pollution.

This approach has the advantage of both providing flexibility and using the same kind of price signals that firms routinely use to monitor, plan and adjust their performance. It does not require a whole army of regulators to be effective. Firms regulate themselves and respond to the competitive behaviour of their rivals in the marketplace (a more detailed analysis of tradeable pollution permits is provided by the Bureau of Industry Economics 1992). These schemes have been tried in the USA with various levels of success. Rosenbaum suggests that the economic benefits of tradeable permits is nowhere near as high as some free-marketeers have hoped for (Rosenbaum 1991: 137).

This approach also identifies an absence of property as the source of environmental problems and proposes a market response. Following Hardin's argument in 'The Tragedy of the Commons' (1968), it is claimed that the lack of private property rights in a common-use situation produces indifference to the long-term economic and ecological health of an asset. Assigning private property rights to the asset generates both revenue to government and a material incentive for an individual or firm to see that the asset survives and is well managed. For example, when elephant numbers have fallen as a result of habitat destruction and poaching and the local population sees wildlife conservation as a threat to its existence, it has been proposed that local people be given property rights in elephants, which they can then use as an economic asset (DiLorenzo 1993). Locals can sell the right to hunt their elephants, sell their tusks for ivory (if the ivory trade is once again permitted) or keep the animals as a tourist attraction. In all cases, locals can make money from their property rights in the elephants. Further, locals now have an economic incentive to care for the elephants and to maintain their numbers. Whereas before, elephants were a dangerous nuisance, damaging crops, fencing and housing, they are now a valuable asset able to return an income to their owners. More than that, locals have an interest in protecting 'their' elephants against poachers. As a result of the granted property rights and associated incentives, it can be anticipated that elephant numbers will rise rather than fall, all other things being equal.

A whole range of these alternative instruments can be applied to a variety of environmental problems, including the application of 'green' taxes to change prices and send signals to encourage good environmental responses to ordinary everyday choices. Nonetheless, such instruments are not appropriate to all problems; some require state action for any chance of success. The best of the 'free-market environmentalists' recognise this and incorporate it into their response.

In the rhetoric of free-market economics, these initiatives are presented as alternatives to state regulation and as a way of avoiding the grip of bureaucratic supervision. At this point, it is important to note that this assertion is wrong. At most, these instruments are 'market-like' rather than market mechanisms. In their design, they harness the price mechanism to pursue designated environmental goals (such as reducing air pollution, improving water quality or preserving stocks of wild life) and they depend for their purpose and efficacy on political decisions and bureaucratic administration. These policy instruments are different ways to achieve the same kinds of goals as direct bureaucratic regulation. In such circumstances, the question of which instrument is preferred is not just a political question but is also a pragmatic one, a choice of means towards politically desired ends. In choosing between them, politicians and officials can make their choice on the grounds of technical efficiency and with due consideration of the kind of problem and the kind of solution sought. It is not a matter of state or market but of the form in which they are combined; it is not a question of regulation or deregulation but of changes in the overall mode of regulation. Even the most libertarian free-market solutions require the allocation and protection of property rights, as well as the policing of contracts and enforcement of agreements.

Consensus and participatory decision-making structures/strategies

One of the most enduring features of environmental politics and policy making is conflict. It does not matter whether the issue is pollution, changed land use, forest clearing, mineral exploration, dam building or quality of life, environmental incidents are frequently attended by groups of environmentalists in conflict with government and private business as they pursue development. Such conflict is completely ubiquitous and it is important. Government responses to environmental concerns are frequently impelled by the conflict that surrounds them. Often it seems as if, without conflict, governments would not respond to environmental problems. Given this, for governments and administration (those making and administering policy) conflict is as much a problem needing response as the environmental conditions to which it is attached.

How do governments respond to this dimension of environmental politics? The answer depends on circumstances, opportunities and the kind of government involved. It also depends what level of government is concerned, be it local, state, federal or international level (see Box 7.2 for local government example). If the government has a strong commitment to development and has either secure control, an authoritarian bent or a strong military base, it may simply ignore or repress dissent and allow development to proceed with whatever forms of environmental assessment are necessary to secure finance for the development project. The building of dams in India, China, Thailand and Malaysia, including projects to generate hydro-electricity on the basis of building a giant dam that displace indigenous populations from forests, are examples.

Box 7.2

Local government and *Agenda 21*

The *Agenda 21* document emerged out of the 1992 United Nations Conference on Environment and Development (the first Rio Earth Summit), and has catalysed activity at various levels of government. One particular chapter encouraged local governments to educate, mobilise and engage their citizens, organisations, and private companies to adopt sustainable development and 'a local *Agenda 21*' (Smith 2002: 6). Local governments have seized upon the Local *Agenda 21* (LA21) policy framework as an opportunity to implement sustainable development through local democracy and participatory decision making, even despite a lack of support and funding at state and national levels (Adams *et al.*). Since 1992, more than 2,000 local governments in 64 countries have established LA21 planning processes (ibid.: 189), while 7,000 towns and cities are implementing the principles of sustainable development.

Adams *et al.* attribute local governments' new pivotal role in environmental policy to three main factors (ibid.: 188), which LA21 builds upon. First, many of the specific issues that derive from international trends like climate change and species loss are local in nature, which has led to a 'strengthening of the connection between global imperatives and local action', also known as 'glocalisation'. Second, there is a recognition that how people live their daily lives within a locality can have significant environmental impacts, and routines of daily life need to be fundamentally changed to prevent this. Third, policy and practice at local government level is itself an important determinant of environmental outcomes.

LA21 did not prescribe any particular process or development, thus there is great diversity in approaches taken around the world. Some of these are more narrow conservation-focused models, while many others address broader urban sustainability issues such as recycling and waste minimisation, air pollution, public transport, traffic management, urban design, stormwater management and energy efficiency. LA21 has also increased local initiatives to address global issues, for example the coalition of over 160 cities that have joined the International Council for Local Environmental Initiatives' Cities for Climate Protection campaign to reduce local greenhouse emissions by up to 20 per cent (ibid.: 190). Yet while being immensely important in implementing local environmental projects and services, most LA21 initiatives so far fall short of addressing the more fundamental Rio questions and implications of injustice and poverty and its human and environmental consequences.

If the government is involved in a liberal-democratic or other effective representational system, its options may be either broader or more narrowly constrained. It is still possible for such governments to choose to ignore environmental protests and opposition. For example, the Thatcher/Major Conservative Party governments had no trouble in using the police to remove protesters and to push through the construction of ring roads and by-passes. In Canada, the government in British Columbia used normal law enforcement to allow logging to proceed when protesters sought to blockade the forests. Governments in the USA and Australia have taken similar actions. Nonetheless, continual public activism and disruptions over environmental matters can be debilitating and politically costly. Under these circumstances, governments may seek to use a whole variety of devices to contain, incorporate or absorb conflict and hence minimise disruption. The US administration, for example, at times creates special 'administrative rules' to determine environmental outcomes (see Box 7.3), although the favourite device of governments of all kinds in many different political systems is the public inquiry. In Britain numerous Royal Commissions and public inquiries have been held on contentious issues that either already had engendered conflict or disruption or had the potential to do so. For example, Royal Commissions were held on the health risks associated with the Windscale Nuclear Power Plant (now called Sellafield), the location and type of reactors for Sizewell B, and for the location of various new airports. Indeed, many inquiries are held at all levels to deal with contentious and conflict-ridden environmental issues. In the United States, inquiries have been used to deal with chemical pollution, contaminated sites and wetland preservation. In Australia, the practice of holding inquiries was made routine through the creation of a Resource Assessment Commission (RAC), which was involved in making recommendations for managing conflicts over uranium mining at Coronation Hill (in the third stage of the expansion of the Kakadu National Park), export wood chipping from native forests and, later, on the management of the multiple uses of the coastal zone. In South Africa, the proposal to sand-mine dunes at Lake St Lucia, a Ramsar convention-listed wetland, was the subject of an extended EIA process that included vast amounts of public consultation and feedback (Preston-Whyte 1995; McEachern 1997).

All public inquiries, regardless of their title or varied modes of operation, have several features in common. They all empower some person(s) to hold an inquiry, to gather evidence (through submissions or by people appearing as witnesses), and to form an assessment on the basis of

Box 7.3

US Administrative rules: from Reagan to George W. Bush

There is a unique situation in the US whereby the lack of lasting environmental administrative infrastructure has enabled incoming governments to reverse or displace past environmental practices through special 'administrative rules'. This is attributed to President Reagan's slashing of spending on the EPA and environment programmes in the 1980s, and implementing an administrative strategy 'to reduce the burden of environmental programs through careful use of the appointment and oversight powers of the American presidency' (Kraft 1995: 410). The use of administrative rules was seen in Chapter 3 with George W. Bush freezing the Cove/Mallard forest protection rule in 2001 that Bill Clinton had previously enacted in 1999. Administrative rules have also been used in the controversial cases of the Healthy Forests Restoration Act of 2003 and the Clear Skies Act of 2003.

The Healthy Forests Initiative was a response to the widespread US forest fires of 2002. The main thrusts of the Act are to thin 'overstocked' stands of forest and clear away vegetation and trees for firebreaks (Healthy Forests website). The Act is opposed by all mainstream environmental groups, such as the Sierra Club, the Wilderness Society and the Natural Resources Defense Council; as they fear that logging companies are allowed to unnecessarily cut large trees under false pretences, and that the acceleration of 'thinning' covers millions of acres which are not at risk of forest fires (Sierra Club 2006b). They argue that logging actually increases fire risk. The official Healthy Forests website depicts a beautiful stand of forest both before and after 'fuel treatment' as part of the 'hazardous fuels reduction project' – the before picture shows a healthy forest full of life, while the after shot is of a forest more or less completely logged with only a thin row of trees left behind (2006). It is a fairly transparent rule designed to favour the forestry industry.

The proposed Clear Skies legislation was sent to the EPA in 2003 after Bush's State of the Union address in which he urged the EPA to pass the measure 'for the good of both our environment and our economy' (EPA 2006). It was basically designed to reduce emissions of sulphur dioxide, nitrogen oxides, and mercury from electric power generation to approximately 70 per cent below 2000 levels (ibid.). It both capped emissions and was an emissions-trading scheme which would supposedly 'deliver certainty and efficiency, achieving environmental protection while supporting economic growth' (ibid.). However it would have weakened the Clean Air Act and resulted in significantly fewer reductions of air pollutants than before, and an overall increase of toxins in the air (Sierra Club 2006a). It is still in a state of negotiation.

evidence and argument. This assessment is then forwarded to some form of governmental authority (a minister, a department, a local council), where the report is published (or suppressed) and the recommendations are either acted upon, amended, applauded, condemned or ignored.

As devices to diffuse or contain conflict, inquiries work in a variety of ways. Inquiries take the ragged and unpredictable forms of conflict and give them order and coherence. They require time, money and research effort to prepare submissions and, sometimes, to take part in the assessment of evidence and rival arguments. Frequently, environmental groups are short of these kinds of resources, and the effort of servicing the inquiry can detract from campaigns elsewhere, even when governments provide subsidies so that these groups can be involved. Further, there is a certain kind of authority that attaches to the 'independent', 'public' or 'expert' inquiry; this gives recommendations some standing that may sanction developments under conditions that would make continued opposition seem at best unreasonable and churlish. At other times, inquiries make findings that can be used to stop developments or impose conditions which make it likely that private developers will move their projects elsewhere. In Australia, the RAC inquiry into mining at Coronation Hill provided the context in which the government could make a decision that prevented mining. With proposals to sand-mine Fraser Island, an inquiry gave reasons for stopping the development. In South Africa, the massive inquiry into a proposal to sand-mine coastal dunes in the Lake St Lucia wetlands, coupled with the political process marking the transition from apartheid to majority rule, provided the context for a decision not to allow that development to go ahead. It is important to recognise that inquiries do not invariably favour approval or disallowal, but the decision to call an inquiry and the fate of its recommendations are always shaped by the political calculations of the government of the day.

Public inquiries and the like have a certain usefulness in dealing with some part of the continual incidence of environmental conflict. Nonetheless, conflict remains endemic to issues of this kind, and it increases the political costs for governments by introducing uncertainty into their political calculations, at least in those countries with open representational systems. Here it is not surprising to find that conflict-resolution mechanisms should be recommended for all kinds of issues as a way of keeping a problem from becoming politically significant. For example, in countries like the USA, Canada and Australia environmental conflict has accompanied the whole history of 'clear-fell' logging and the wood chipping of old-growth forests. In this context, it is not surprising to find attempts to discover cheaper and routine ways of trying to contain/diffuse conflict, largely on the basis of the sharing of knowledge. One of the most explicit cases of this can be found in the ESD process in Australia. As discussed earlier, the ecologically sustainable development process in Australia was a model of an inquiry that sought to incorporate the destabilising pressures from business, labour and some environmentalists.[4] In the working party on the forests (which did not include environmental representatives), a whole series of recommendations were made on ways of institutionalising and routinising forest conflict, taking it out of the forests and into information-sharing venues. Computer models, roundtable forums, information sharing, forest tours and mediation processes were all recommended. It is assumed that if information is shared and all sides have a say then the chance of significant (politically disruptive) conflict is reduced and that old-growth forests can continue to be clear-fell logged in an ecologically satisfactory way.

Sometimes, such conflict-resolution procedures may help to contain conflict and, if they are effective, the views of environmentalists can be harnessed to the process of making a development project both more environmentally acceptable and more efficient. In these cases, the knowledge of environmentalists becomes harnessed to the process of enhanced, effective and politically acceptable development. Of course, such techniques cannot be effective all the time, in all places and for all kinds of environmental conflict, but the urge to find such devices will often be there.

The citizens' inquiry

On the other occasions, environmentalists have sought to mimic the state's administrative processes in a bid to harass the state within their own institutional frameworks. Various forms of public participation outside the administrative systems of the state have emerged in North America, parts of Europe and Australia in recent years, such as *people's inquiries*, *citizens' juries* and *citizens' forums*. Hendriks describes the basic characteristics of these more deliberative democratic processes as follows:

> While there are some differences between these processes, they seek to bring together a small panel of randomly selected lay citizens to deliberate on a policy issue. After hearing from, and questioning a

> number of experts such as academics and interest groups, the citizen's panel develops a set of written recommendations. This document then feeds into the policy process either directly (for example, tabled in parliament) or indirectly through wide public dissemination.[5]

People's inquiries are different to public inquiries constructed by the state in the sense that there is an attempt to get citizens involved in the policy process at the outset, thus providing for an early deliberative democratic input by laypersons; rather than just making submissions or otherwise responding to the findings of policy-making elites. In this sense, people's inquiries, juries and forums are attempts to 'get democracy in early' (Dryzek 1992: 8).

A public inquiry into uranium was held in South Australia in 1997–8 as a way of dealing with a new Federal government which was deemed by environmentalists to be pro-nuclear and anti-environment (Doyle 2005). A senate select committee into uranium mining and milling had been called previously by federal parliament in 1996, to open the legislative gates for the nuclear industry while appearing to be open to dissenting public opinion. However even before the inquiry began, the then Minister for the Environment, Senator Robert Hill, had stated that the inquiry would be of limited value and would not affect his environmental assessment of new mine applications (*The Australian*, 19 April, 1996). According to the Nuclear Issues Coalition (NIC), the senate inquiry, critics argued, was merely a means of fast-tracking the state/industry objectives, and any promotion of public input was meant as no more than political window-dressing. This was evidenced by the fact that the inquiry hearings were never publicly advertised, and that many environmental organisations who made submissions and requested a hearing were never allowed to give evidence.

Within this context the public inquiry into uranium emerged, sponsored by a large array of groups and organisations, including minor political parties, church groups, women's groups, peace groups and a plethora of environmental interests. This community alliance set up its own roundtable, choosing to mimic the inquiry mechanism of the state in its bid to pursue its goals. But unlike the narrower agenda of the senate inquiry, the public inquiry refused to isolate mining activities from nuclear weapons; the treatment and storage of nuclear waste; and the development of alternative energy options.

The public inquiry was a hybrid mixture of the more 'democratically pure' citizens' juries and the more traditional senate committee inquiry which had preceded it. All 'stakeholders' were invited to take part, but on this occasion the agenda and minimal rules for entry had been established by environmentalists and other community activists – not by the state or extractive industries.

This broader context of policy-relevant inquiries and forums signals some notable experimentation with deliberative processes that poses some challenge to what Deborah Stone has called the 'rationality project' of conventional approaches to public policy. There is a move away from the conventional reliance upon analyses by established experts towards forms of deliberation involving citizens. This move also creates an opening to throw into question the market model of society that is largely taken for granted by disciples of the rationality project:

> Society is viewed as a collection of autonomous, rational decision makers who have no community life. Their interactions consist entirely of trading with one another to maximize their individual well being. They each have objectives and preferences, they each compare alternative ways of attaining their objectives, and they each choose the way that yields the most satisfaction. The market model and the rational decision-making model are thus very closely related.[6]

The very process of deliberation by citizens implicitly opposes the premise that 'there is no such thing as society' (as Margaret Thatcher once put it). Citizen deliberation is, indeed, based on the premise that it is possible for a democratic citizenry to debate and weigh issues in terms of a meaningful public interest. However, it is common for such deliberative experiments to remain linked to the state in a way that makes them ineffective as citizen forums. The powerful intersection of state and business interests typically screens out or marginalises divergent perspectives.

The public inquiry into uranium arose to directly counter the dominance of such forces in Australia by encouraging the public expression of marginalised voices, particularly those of environmentalists. Significantly, these voices did not call for alternative green 'ways of knowing', but demanded rationality and objective science, including the challenges that can come from ecological analyses. The implicit point here is that the conventionally approved rationality project (as Stone terms it) is itself impossible when the state abandons a commitment to independent science and allows business forces to dominate the policy process (se next section). Too often, these forces are interested only in the kind of narrowly focused research and analysis that is needed to maintain and expand their operations. At the same time, whenever inconvenient or troublesome information is encountered, it is typically ignored or suppressed. What is obviously lacking is the kind of comprehensive monitoring of business ventures and their impact that would allow for the meaningful use of ecological science to protect the environment.

Australian environmentalists certainly had not come to endorse the rationality project, but their demands at the public inquiry implicitly suggested that the project is clearly incoherent in a context where the policy process is biased in favor of developments promoted by state and business interests. Here any rationality depends on confronting the conventional orientation with divergent perspectives that are interested in a more thorough comprehension and control of what is done to the environment. This does not mean departing from science, but promoting scientific activity that is independent enough of dominant forces to seek out troublesome information and to analyse what is being done to the environment, not in fragmented terms, but in a manner that is informed by the 'subversive' science of ecology. This kind of independence cannot be gained simply by adhering to the rationality project's call for expert analysis, but depends on the loud and dramatic entry into public debate of environmentally informed voices, uncontrolled by the state and unintimidated by business.

It was thus not surprising that the public inquiry did not attract any corporate input. Also, only one government agency, the Office of the Supervising Scientist (Environment Australia), made a submission and presented evidence at the hearings. In letters to the public inquiry, both corporations and state officials explained their absence by referring to their past contributions to the earlier senate committee inquiry. The refusal of corporations and other powerful interest groups to operate outside their own forums and their own terms of reference, however, is a usual occurrence in deliberative, democratic forums, when their control is limited (Dienel and Renn 1995: 127–28).

The public inquiry into uranium proved to be a tremendous educational tool because information about the uranium industry was – and remains – highly inaccessible. Although it concentrated on WMC's operation at Roxby Downs, the inquiry also adopted a broad focus on gathering evidence on industry operations relating to the full nuclear fuel cycle, to health and safety issues, and to control of information.[7] The inquiry also fostered solidarity among community groups, something that is important because this is often the only source of power these participants possess. An excellent form of theatre, moreover, the inquiry attracted good media coverage.[8] Along with numerous ethical questions, finally, issues of information and scientific 'objectivity' and of the usefulness of deliberative forums re-emerged in the public realm.

The public inquiry into uranium directly encouraged the public expression of marginalised voices, particularly those of environmentalists. The public inquiry was radical, part of a longer-term environmental struggle. As a model for public education and symbolic politics, it most importantly showed the potential for public deliberation as a form of action confronting the powers that be (Doyle 2005b). Interestingly, the Howard Government launched a new, hand-picked 'Nuclear Energy Taskforce' (the Switkowski Review) in 2006 in an effort to legitimate the nuclear industry's capacity to provide commercially viable, climate friendly energy by 2020. In response to their perceived 'lock-out', dissenting anti-nuclear communities have again created another citizens' enquiry based on the earlier proto-type.

Public service reform and the role of independent science

Since the 1990s there has been significant change in the form and practice of public administration. From a Weberian model of the efficient bureaucracy, the more popular view of bureaucracy as costly, slow and inefficient has dominated. The image of the market and the private sector is in the ascendant and most countries have reformed their public sectors to mimic the perceived benefits of private sector models. The size and scope of the public sector has been reduced. Privatisation, the sale of public assets to the private sector, has been significant. Contracting-out activities previously carried out by state administration has increased dramatically. Where state assets have not been sold, frequent recourse has been made to 'corporatisation' to get the remaining agencies to behave as if they were private sector bodies generating a return on their investments. Purchaser–provider models, which had their origins in the reform of public sector health provision, have spread to other areas of government activity. In general there has been a sustained change in the role and purpose of regulation. Market-like instruments and self-regulation of business are increasingly preferred over direct regulation.

How have such public sector reforms affected the role of the administration and administrative agencies in the regulation of the environment? Environment protection agencies have not been immune from the pressures for public sector reform. In the past there have been attempts to make such agencies assess the costs and benefits of their regulations. This was especially true with the American EPA in the Reagan era. Such environmental agencies have been downsized, restructured, had their range of responsibilities redrawn and initiatives have passed to consultants and the private sector. Since the rise of environmental concern, business has never been subject to less direct regulation of its activities.

The case of the state of South Australia is significant here, with the state often presenting itself openly as a corporation, retreating from its past social and environmental responsibilities. Both the South Australian Department of Primary Industry (PIRSA) and the Department of Environment, Housing and Aboriginal Affairs (DEHAA) have undergone a structural transition to a purchaser/provider model (Doyle 2000: chs 11 and 12), which has been echoed in numerous other bureaucracies under free-market governments elsewhere in Australia and the world. The purchaser/provider model dictates that the purchaser, or client, is no longer the public. Each bureaucratic division's client is now its own departmental director and his or her policy advisers. The director purchases the services from his or her own employees. When these services can be more efficiently and effectively purchased outside of the bureaucracy, the director outsources them. Where once the public would demand that the public service act in its interests, now the bureaucracy must satisfy its own top managers, who, in South Australia, are sometimes political appointees of a party whose interests are often inseparable from those of big business. The bureaucracy must now provide the 'correct' information that will lead to profitable outcomes. In effect, business becomes the client and government becomes its service provider.

The attack on traditional public service has also been accompanied by a thorough offensive against independent science. In many ways independent science complements the concept of a separate, permanent state, performing a monitoring and moderating role in the processes of capital production, accumulation and consumption. Whatever the limitations of bureaucracy and positivist science, both have provided an alternative force to the excesses of capitalism. Just as an autonomous state is often seen as interfering in radical free-market environmental solutions, 'independent' science has also been seen as a barrier to market-based outcomes.

The funding and political support for the Commonwealth Science Industry Research Organization (CSIRO), traditionally Australia's largest state-run science organisation, has been dramatically slashed in real terms under the Howard government's neo-liberal economic regime. Generally, however, the attack on science is more subtle, masked in the changes to the philosophy of government already alluded to in the discussion of the purchaser/provider model of the public service. Fewer permanent scientific officers are being employed, with more and more services being outsourced to industry-friendly consultants. It is an open question whether these new forms of supervising the impact which society has on its environment are more or less effective than the old versions.

International administrative institutions? Diplomacy, conferences, conventions and regimes

There is no world government with a permananent bureaucracy governing our lives. Increasingly, however, through international diplomacy, there is an ever increasing list of administrative regimes and conventions that are emerging which influence if not actually govern our global environments. There are three ways in which environmental problems raise the need for international institutional responses. First, there are problems shared by a large number of countries across the globe. Questions of pollution, soil degradation and forest depletion are of this kind. Since these problems are shared, there is some point in holding international conferences and agreeing on broad policy frameworks. Much of what happened at Stockholm in 1972, in the production of the World Conservation Strategy in 1980 and the Brundtland Report of 1987, focused on sharing an appreciation of common problems and the array of policy instruments that can be used to respond – a sharing of awareness and experience. Second, there are problems that may be concentrated in some countries but which carry global implications. Energy and resource use in the United States (as the largest consumer) carries such implications, since it acts as a constraint on the standard of living in other countries and limits their ability to have access to equivalent quantities of resources at similar prices. Population issues also fall into this category. Again, these issues can be addressed in global forums but they cannot plausibly be solved there. Third, there are those issues that are either cross-border in character or are genuinely global. In the case of cross-border issues, diplomacy and the resultant international policy-making are required. Genuinely global problems require genuinely global solutions.

This chapter ends with a consideration of three diverse global policy-making initiatives: the Rio Earth Summits of 1992 and 2002; and the Basel Convention of Hazardous Wastes; and the emergence of the World Trade Organisation.

The Rio Earth Summit and *Agenda 21*

Numerous global conferences have been organised around environmental themes since the first in 1972 in Stockholm. Many of these have been associated with the United Nations Environment Programme (UNEP) as it has provided personnel, funding, information and education to support these global initiatives. The Stockholm conference was, in a very real sense, a preliminary affair (Ward and Dubos 1972). It was more involved in promoting awareness and outlining a timetable of activities (which culminated in a big environmental conference in 1992) than in drafting and issuing some broad policy framework in the shape of the usual communiqué. UNEP was subsequently involved in the process that produced the World Conservation Strategy (1980) and the Brundtland Report (1987), as well as promoting the research and diplomatic initiatives that eventually produced the Montreal Protocol on ozone-depleting substances. It has been a major player in seeking and promoting both global awareness of environmental issues and global responses to environmental problems. The 1992 conference was to be its major achievement.

The first Earth Summit, or more formally the United Nations Conference on Environment and Development (UNCED), was held in Rio de Janeiro in Brazil from 3 to 14 June 1992. Jordan writes:

> It is difficult not to review Rio without repeating some of the 'key' (but now hackneyed) facts: 'the Summit was attended by 130 heads of state, 1500 non-governmental groups and 7000 accredited journalists'; 'it was the largest high-level intergovernmental conference ever held'. But, 'Rio' was only the most visible tip of a much bigger iceberg of activity. Prior to the Summit, there had been almost two and a half years of international organisation(s) at a variety of fora dotted across the globe: preparatory committees; intergovernmental negotiating committees; meetings of the G77 (developing) states. Eventually, these negotiations culminated in the agreement of two opaquely worded 'hard law' conventions on climate change and biodiversity protection, and three pieces of international 'soft law'.
>
> (1994: 160)

Some degree of institutional redesign was also accomplished at Rio with a 'reconfiguration' of the World Bank's Global Environmental Facility (GEF) and the establishment of the UN Commission on Sustainable Development (UNCSD). The largest document released at Rio was *Agenda 21*, an account of what needed to be done to secure a global future into the twenty-first century. Rather like the documents that went before it (the *World Conservation Strategy*, *Our Common Future* and the initial Stockholm conference proceedings), *Agenda 21* combines an account of environmental and ecological damage with a suggested set of priorities and policy responses to achieve some form of sustainable development. Producing *Agenda 21* was an enormous and frustrating exercise in drafting, negotiating and redrafting to find an acceptable version that could be authorised by the Earth Summit. Indicating the character of this compromise and consensus is the first principle, which begins: 'Human beings are at the centre of concerns for sustainable development' (UNCED 1992).

The politics of the Rio Earth Summit are worth evaluating. There was a sense in which the conference and *Agenda 21* reinforced understandings of the global dimension of many environmental issues, and this was reflected in the popularity of terms like 'interconnectedness', which were freely added to the public vocabulary. It was broadly accepted that the environmental problems of a country or region can have an impact, either directly or indirectly, on the rest of the inhabitants of the globe.

This recognition of interconnectedness and the globally shared character of environmental problems influenced the policy initiatives of the governments of the more affluent countries. In the past, some environmentalists like Ehrlich, some deep ecologists and the neo-Malthusians defined population pressure as the key global environmental issue and a threat to their parts of the world. They sought to defend their environments by separating out their countries from any global context and sought to confine the consequences of population growth to the poverty of people in remote parts of the globe. To the extent that they had a global perspective, it was to pressure poorer countries to take efficient methods to lower the birth rate. A recognition that the biosphere had to be shared, that the demands of economic growth in one part could endanger environmental conditions in another, forced a rethinking of this perspective. Some environmental activists and some governments sought to use these global forums as a way of pressing for rules of conduct that would protect their countries from the harmful consequences of bad practice in other parts of the world.

The prescriptions for sustainable development can be interpreted in this way. Although *Agenda 21* sets out guidelines for achieving a form of global environmental protection, the negotiations that produced it marginalised a number of NGOs and excluded a whole range of issues.

Much was made at the conference of the inability of government and, particularly, non-governmental representatives to address substantive issues that they had deemed important. Wagaki Mwangi, a representative of the Nairobi-based International Youth Environment and Development, expressed these complaints in this way:

> Those of us who have watched the process have said that UNCED has failed. As youth we beg to differ. Multinational corporations, the United States, Japan, the World Bank, the International Monetary Fund have got away with what they always wanted, carving out a better and more comfortable future for themselves . . . UNCED has ensured increased domination by those who already have power. Worse still it has robbed the poor of the little power that they had. It has made them victims of a market economy that has thus far threatened our planet. Amidst elaborate cocktails, travelling and partying, few negotiators realised how critical their decisions are to our generation. By failing to address such fundamental issues as militarism, regulation of transnational corporations, democratisation of the international aid agencies and inequitable terms of trade, my generation has been damned.
>
> (quoted in Chatterjee and Finger 1994: 167)

Why was it that the vast array of delegates, official and unofficial, who gathered at Rio had so little impact on what happened and the content of the documents issued? Earth Summits, even where they involve a mass of delegates, are just like all other summits held between world leaders. What is achieved, signed or published is usually the product of extensive diplomatic negotiations, which take place behind closed doors and before the conference is held. At most, summit talks clarify and resolve a limited set of outstanding, difficult or sensitive issues. The success or failure of UNCED had very little to do with what actually transpired at Rio; the broad details of what was to be agreed had already been determined by the more powerful states attending, the conference secretariat and powerful business organisations. The conference itself was largely the 'show' used to 'launch' the product, *Agenda 21* and the various conventions and agreements.

The United States largely shunned the Rio process and the documents that were born there. Under the leadership of President George Bush, the

United States was largely interested in promoting increased protection of tropical rainforests, while at the same time seeking to weaken agreements on climate change and biodiversity in its own favour. In fact, the USA was the only industrialised country not to sign the biodiversity convention. The US stance was not determined by the politics of the Earth Summit and diplomatic negotiations between countries, but by the internal dynamics of US domestic politics and White House political calculations. Schrader writes:

> To many participants, Bush seemed paralyzed by the dominant impulses of his Republican party. The internal battle over the treaty had been fought and decided months before the conference. David McIntosh, the executive director of the President's Council on Competitiveness . . . had ruined the Rio treaty and the EPA's chances for success by savaging the policy in an April memo. He warned that the treaty would impair the ability of American corporations to protect their biological inventions and technologies overseas.
>
> (1992: 431)

Bush's concern with his image of 'toughness' and 'leadership', the up-coming presidential election and the fractious nature of the Republican Party shaped the way in which the USA both supported and limited initiatives taken at Rio. Once Bill Clinton had been sworn in as president, domestic considerations were again behind his decision to sign the Biodiversity Convention. In practice, the change was no more than symbolic. There was no change in the way in which the USA defined its interests on these global environmental issues, just a change in the ways in which those interests would be presented and defended.

All international summits and conferences are full of hype. Expectations of dramatic happenings and outcomes are created, only to be disappointed by the mundane proceedings. For a time, the world's media came to be focused on the environment. Environmental activists and government players gathered together, networks of contacts were created and the conference closed. The circus moved on. Rio was an overblown affair (both in terms of costs and expectations), and its achievements were modest and could have been attained more cheaply by conventional means. As a result, conferences of this kind became relatively discredited. The set agenda of mega-conferences would continue (the Population Conference in Cairo, the Women's Conference in Beijing and the Habitat Conference in Istanbul), but it was unlikely that an Earth Summit like that in Rio would be held again. Sandbrook writes:

> The world has learnt that there is no easy set of solutions to be had at an international level to many of the world's environment and development ills. National and subnational administrations hold the key. If they respond to this challenge, there will be progress . . . The optimists would argue that there is now on the table a new and powerful agenda relating to long-term welfare and a protected planet. The question is who will pick it up and do something about it.
>
> (1993: 30)

The spirit of Rio has not come to an end, despite the time that has passed since the first conference was held. Since then, attempts have been made to complete the drafting of the Earth Charter. Maurice Strong, the driving force behind both the promotion of sustainable development and much environmental summitry, has combined with US philanthropists, including the Rockefeller Foundation. The preamble to the draft Earth Charter gives some senses of its perspective:

> As a global civilisation comes into being, we can renew efforts to build a truly democratic world . . . These aspirations can only be fulfilled if each and every person acquires an awareness of global interdependence and decides to live according to a sense of universal responsibility.
>
> (Earth Charter Drafting Team 1999)

Conferences and meetings are being held across the world, seeking to get commitments to this approach, urging support for the Charter as a way of giving the market a global soul. A visit to the campaign's website at http:www.earthcharter.org reveals various resources which illustrate the approach and the thinking behind it. As could be expected, the major focus of the Earth Charter campaign is on value change with little attention to the realities of inequality and the unequal distribution of power between the richer and poorer parts of the Earth.

Rio illustrates that the benefits of such conferences are difficult to measure. Apart from the 'soft' agreements and the brief but intense focus of the international media, delegates (both governmental and non-governmental) established networks of contacts that live on after the event. In addition, each conference manages to take on board major criticisms of the previous event. For example, 'Habitat 2' at Istanbul in 1996 did manage to address the impact of trade liberalisation on poorer communities, particularly with reference to issues of shelter.

Ten years later, the 'Rio + 10' Earth Summit was held in Johannesburg, South Africa, also known as the World Summit on Sustainable

Development. The Johannesburg Declaration was the main outcome of the summit, which included agreements to restore the world's depleted fisheries and halve the number of people who do not have access to basic sanitation by 2015. However, according to Friends of the Earth International, these were the only two new targets while all other existing commitments were simply 'reaffirmed, watered down, or trashed altogether' (FoE International 2002). The absence of the US through a public boycott by George W. Bush diminished the potential of the summit. The summit was also criticised for the hypocrisy in serving expensive catered food and drink for dignitaries while South Africans starved nearby. Certain instrumental organisations and individuals were excluded, especially those critics who believed that 'sustainable development' was an attempt to greenwash economic development at the expense of long-term environmental goals.

International environmental politics, as practised in the series of United Nations-inspired conferences (like Rio), have been both effectual and ineffectual. Such conferences are good for producing formal statements, but they often fail to establish international institutions to monitor and regulate global environments. *Agenda 21* did produce, however, a detailed administrative framework which has been voluntarily adopted in some countries to guide bueaucratic procedures (see Box 7.2).

Hazardous and radioactive waste disposal: the Basel Convention

> No universal definition exists of 'waste' and national definitions vary considerably and are often subjectively based. One legal definition is that waste comprises 'substances or materials which are disposed of, or are intended to be disposed of, or are required to be disposed of by the provisions of national law'. 'Hazardous waste', on the other hand, consists of materials of varying categories and having specific characteristics that are generally considered harmful to human health.
>
> (Bamako Convention, article I, annex 1 and annex II)

For the most part, the transboundary movement of hazardous waste has followed the 'path of least resistance' principle. Furthermore, there is generally an 'economic and regulatory imbalance' between the generating and the importing states. Thus, dumping at sea is easier and less costly, and the developing world, and particularly Africa, has been seen as an attractive recipient region on account of relatively inexpensive

Plate 14 ***Ecological Debt banner by Acción Ecológica (Friends of the Earth, Ecuador) at the 2004 World Social Forum, Mumbai, India. Courtesy of Joel Catchlove***

disposal costs (Krummer, 1999, 4–7; Lipman, 2002). However, post-Cold War international law, especially in the form of the Basel Convention, has been developed in order to regulate the transboundary movement of hazardous waste.

The Basel Convention on the Control of Transboundary Movements of Hazardous Wastes and their Disposal came into force in 1992. One of the more important concerns lying behind the Convention was the fact that, of the more than 400 million tons of hazardous waste produced each year, approximately 80 per cent came from OECD countries and a large proportion was being 'dumped' in industrialising states (Greenpeace 1998). The scope of the Basel Convention, however, does not extend to the transboundary movement of *radioactive* materials, although it does allow some transboundary movement of hazardous waste 'only when the transport and the ultimate disposal of such wastes is environmentally sound'. This clause has been used as a bargaining position for those states possessing hazardous waste exports since it could imply unlimited external access so long as appropriate local environmental expertise and technology existed. An important amendment to the Basel Convention, however, adopted in 1995, prohibited the export of all hazardous wastes from richer to poorer states from 1998 (Rumley and Doyle 2005).

The issue of the transboundary movement of nuclear waste was explicitly not inserted within the Basel Convention because its movement was seen as falling within the sphere of competence of the International Atomic Energy Agency (IAEA). The adoption in 1990 by the General Conference of the IAEA of a non-binding code of practice on the International Transboundary Movement of Radioactive Waste, which was proposed by UNEP, however, has in fact resulted in a weakening of the principles underpinning the Basel Convention (Krummer 1999: 85).

The disposal of hazardous substances is also an increasing bone of contention in the global sphere, given the transnational nature of the nuclear cycle which involves mining, milling, enrichment, reactor use, reprocessing and waste disposal (not to mention nuclear weapons fabrication). Key here is the transport of nuclear waste which crosses both state, national and international boundaries. The two main areas of concern are radioactive waste being transported by sea and land, which includes both mining wastes and the more toxic substances which evolve after the processes of nuclear reaction. Here we see interlinkages and conflicts between energy security, military security and environmental security.

While Africa possessed the dubious distinction of being 'first choice' for the dumping of European nuclear waste and persistent organic pollutants (POPs) in the 1980s, it was the first to respond politically to the threat of 'waste colonialism' (Bernstorff and Stairs 2001: 4). Prior to the ratification of the Basel Convention, many African states were especially concerned about the transboundary movement of such hazardous waste into Africa from industrialised countries, and some indeed saw this process as one of the 'dumping' of nuclear waste into Africa (Krummer 1999: 99). At the May 1988 Organisation of African Unity (OAU) Council of Ministers 48th Ordinary Session in Ethiopia, a resolution condemned the importation into Africa of industrial and nuclear waste as a 'crime against Africa and the African people' and called upon member states to introduce import bans. The resolution condemned 'all transnational corporations and enterprises involved in the introduction, in any form, of nuclear and industrial wastes in Africa; and DEMANDS that they clean up the areas that have already been contaminated by them' (Organisation of African Unity Secretariat, CM/Res.1147–1176, 1988).

As a consequence of this resolution, work began on an African Convention under the auspices of the OAU shortly after the adoption of the Basel Convention, since the latter excluded nuclear waste. There was therefore a concern that certain of the needs of African states were not properly taken into account, and thus, while the Basel Convention was a Convention of the North, there was need for a Convention of the South. The resultant Convention, which was adopted in Mali in January 1991, entered into force in April 1998 (Bamako Convention). However, adherence and compliance are major problems, with many states most affected not signing or ratifying the convention (Rumley 2005: 14). Like the Kyoto Protocol, nation-states usually retain the power to decide whether or not to enter into international regimes and conventions. Like the Rio conferences previously addressed, these conventions provide guidelines for environmental behaviour, rather than making the transition into strict international laws which can be policed in a manner which alters the policy making of institutions within nation-states.

The rise of the World Trade Organisation: global institutions with power to enforce

Certainly, Rio was important but not as important as the protracted negotiations that produced the World Trade Organisation (WTO) and the new General Agreement on Tariffs and Trade (GATT). GATT has clauses that limit the circumstances in which environmental regulation can be imposed if, by acting, there is an impact on trade. It is possible for a country to regulate business interactions with the environment, but it is not possible for it to do so in such a way that it has an impact on trade. The government of one country can challenge the validity of environmental regulations and restrictions in another country and have them rejected. The acceptance of the 'environment' as an issue by the World Bank and the International Monetary Fund (IMF) is to be understood in these terms, as the product of the changes in the institutional setting, partly as a result of the United Nations Conference on Environment and Development and partly following the creation of the World Trade Organisation.

The rise of the World Trade Organisation (WTO), formed in 1995, has significantly changed the arena within which environmental concern is expressed on a global scale. The WTO, the culmination of GATT (General Agreement on Tariffs and Trade) negotiations, has changed the ways in which countries can resist attempts to impose environmental concerns on others. The WTO exists to promote free trade and the dismantling of all barriers to free trade, including environmental concern. Under WTO procedures states are restricted in the measures they can take to promote their environmental goals, either domestically or in their trade relations with other states. For example, the WTO can impose trade sanctions to prevent a country boycotting shrimp caught without the sea turtle excluder devices that would reduce the kill of this endangered species (2002: 283). Subsidies and other government funding to promote environmental best practice may be challenged by trade rivals as distorting markets and hence can be punished by retaliatory sanctions. Governments cannot put restrictions on trade on environmental grounds. Of course, the WTO does not see itself in this light. A visit to the WTO web site (http://www.wto.org/) provides a clear account of this defence against such criticisms. The WTO sees no tension between its goal of improving world trade and countries caring for the environment. Most of its defence turns around the standard appeals to the efficiency of the market, assuring the best allocation of scarce resources which must, by definition, be environmentally sound. Indeed in some areas this claim can be correct as with irrigation subsidies which do not take into account the environmental costs associated with such a use of water (not that there is any evidence that the WTO has been effective on this question). In other areas it is clear that WTO rules and rulings will seek to limit expressions of global environmental concern. The WTO, even in its defensive mode, notes this when it states: 'Also important is the fact that it's not the WTO's job to set the international rules for environmental protection. That's the task of the environmental agencies and conventions.' Elliott (1998: 207–14) gives a clear exposition of the tensions between freer trade and a better environment and outlines what would be required to establish a harmony between the two objectives. One of the clear areas of tension concerns biotechnology and the production of food. European countries have been less enthusiastic than their US and Canadian counterparts over the sale of hormone-fed beef and genetically modified food. Hormone-fed beef has already been subject to WTO determination to ensure its access to European markets, regardless of the wishes of European governments. Much the same is looming in the case of genetically modified food. Policy makers in the USA have stated that they intend to take the Europeans to the WTO over their requirement that GMO food should be labelled – which is seen as a restraint of trade and not related to issues of food safety or legitimate environmental concern.

Partly because of its ambiguous claims to be environmentally friendly, the WTO has been subject to increasing pressure to change its rules and procedures by a diverse coalition of environment and labour organisations. The street demonstrations during the 1999 Seattle meeting of the WTO illustrated some of this opposition. Since then the WTO has sought to define new rules for dealing with restrictions in trade based on environmental grounds. This included a decision at the WTO Ministerial Meeting in Doha in 2001 to give greater attention to the subject of multilateral environmental agreements which have emerged since Rio. However, the WTO has expressed interest in overturning the trade-related provisions through its own dispute settlement procedures, which would again undermine the autonomy of these multilateral environmental agreements (O'Brien and Williams 2004: 312). It will be necessary to wait and see if these proposed changes make any difference to the expression of environmental concern or not.

The WTO was supposed to be complemented by a parallel agreement on investment, the Multilateral Agreement on Investment (MAI). This too would have restricted the ability of states to impose restrictions and conditions on environmental grounds. The MAI was temporarily stopped in the wake of widespread, internet co-ordinated, petitions and actions by a whole variety of groups from all over the political spectrum. The MAI initiative is not dead. Attempts to negotiate an agreement will resurface in the coming years.

Yet there is a flipside to these new forms of global environmental governance in the myriad of emancipatory NGOs, groups and individuals that form the global justice movement (see Chapter 3). The movement is becoming increasingly organised while remaining decentralised, and hosts the annual World Social Forum to directly counteract the World Economic Forum meetings in Switzerland (see Box 7.4).

Conclusion

The politics of bureaucracy, the institutional and legal design of the state, and the variety of policy instruments chosen to respond to both problems in the environment and increasing levels of environmental concern vary from country to country and from time to time. Changes in design and preferred policy instruments do not seem to be a product of increased skill in the making of environmental policy or a result of increased awareness of the interconnection between economic development and environmental consequences. Rather, the structure of institutions, the procedures for their co-ordination and the choice of policy instruments seem to be more influenced by a broader debate about the character of

Box 7.4

World Social Forum

Since 2001, the World Social Forum has been an annual open meeting place where social movements, networks, NGOs and other civil society organisations opposed to neo-liberalism and a world dominated by capital or by any form of imperialism, come together to pursue their thinking, to debate ideas democratically, to formulate proposals, share their experiences freely, and network for effective action. The WSF is guided by a Charter of Principles which enshrines its plurality and diversity, and its non-confessional, non-governmental, and non-party nature. It proposes to facilitate decentralised coordination and networking among organisations engaged in concrete action towards building another world, at any level from the local to the international, rather than be a body representing world civil society.

Meetings so far have been held in Porto Alegre (Brazil), Mumbai (India) and the more recent polycentric meeting was held in Bamako, Caracas, and Karachi, more or less simultaneously. It tends to meet at the same time as its capitalist rival – the World Economic Forum in Davos, Switzerland. The WSF has now prompted the organising of many regional, local, and national social fora, most of which also adhere to the guiding principles. It is the global justice movement's way of building viable alternatives to what it sees as the current, unsustainable world order.

Source: Adapted from *World Social Forum: Another World Is Possible* website, http://www.forumsocialmundial.org.br

government and its impact on economic growth. Hence, when the mood changed against state ownership, regulation and intervention, there was a corresponding change towards market-like policy instruments. No matter how strong the preference is for markets, governments and bureaucracies – from local to global – set the frame within which environmental policies are made and evaluated.

The basic lessons of the three examples of attempts to create global institutional and administrative responses reveal the limitations inherent in a system of nation-states responding to environmental problems that are shared, problems that cross national boundaries and those that are truly global. There is plenty of scope for global forums where world leaders meet and NGOs gather to discuss shared problems and exchange views on possible policy instruments and experience. Such forums are the focus for the preparation of major statements, such as *Agenda 21*, *Our Common Future* and the occasional protocol. Global forums and diplomacy can produce solutions or the frameworks for addressing some environmental and trade issues, such as transboundary hazardous waste, but there is no simple blueprint for success. Some problems, like global warming and the control and treatment of environmental refugees, test the limits of the ability to respond. The most enduring challenge for environmental politics and policy making lies at this international level.

Further reading

Chatterjee, P. and Finger, M. (1994) *The Earth Brokers: Power, Politics and World Development*, Routledge, London and New York.

Dryzek, J. (1987) *Rational Ecology: Environment and Political Economy*, Basil Blackwell, Oxford.

Dryzek, J. (1997, 2005) *The Politics of the Earth: Environmental Discourses*, Oxford University Press, Oxford.

Kütting, G. (2000) *Environment, Society and International Relations: Towards More Effective International Environmental Agreements*, Routledge, London and New York.

O'Brien, R. and Williams, M. (2004) 'Governing the Global Political Economy', Chapter 11 of *Dynamics*, Palgrave Macmillan, New York, pp. 287–314.

Paehlke, R. and Torgerson, D. (eds) (2005) *Managing Leviathan: Environmental Politics and the Administrative State*, Broadview Press, Peterborough, Canada.

Pepper, D. (1984) *The Roots of Modern Environmentalism*, Croom Helm, London.

Conclusion: environment and politics

In this book, we have provided an introduction to understanding the comparative dimension of environmental policy and policy making. Environmental problems are either shared between, or common to, a large number of countries, so it is possible to consider the similarities and differences between their responses. To make an effective comparison it is necessary to describe what is being done in response to environmental challenges and to note variations between countries and over time. It is also necessary to have a framework within which the comparison can be evaluated. In this book, we have concentrated on providing a series of alternative ways of analysing and evaluating what happens in different countries and at different times. We have been concerned to give accounts of the different concepts that are necessary for the comparative enterprise, as well as the different arguments that are made both about these concepts and the ways in which these interpret responses to environmental problems. In addition, we have illustrated our presentation with examples drawn from a wide range of countries in quite different situations. For instance, we have drawn examples from the United States of America, Britain, Western Europe and Australia to illustrate the response in relatively wealthy, well-industrialised countries with liberal-democratic political systems. Contrasts have been sought from the countries of Eastern Europe, which experienced environmental degradation as a result of state action pursuing rapid industrialisation, and from countries like Iran and China, which are caught between an urgent need for economic development and concern about its environmental consequences.

Central to the presentation of the opportunity for a comparative assessment of environmental politics is a concern with its necessary interdisciplinary character, bringing together in a single frame the contributions of a wide range of disciplines of both the physical and social sciences. A series of core concepts have been discussed. The character of the political system, from liberal-democratic through to authoritarian rule, has an important impact on what is done (or not done) in the name of environmental concern. The concept of power, from behaviourists and pluralists through to the neo-pluralists and Foucauldians, is central to an understanding of what is done, why it is done and how these actions are to be interpreted.

From a consideration of the core concepts of political regimes and power, the discussion moved on to an account of the wellsprings of environmental action: the range of arguments, claims and counter-claims that impel different responses to environmental problems. For example, some claim that nothing needs to be done, since environmental problems are either not serious or are easily solved. Others note environmental problems and propose reforms but seek to act within the prevailing assumptions of economic growth. There are still others who have developed more radical accounts of what causes environmental damage and what cure is needed. This last group has formed the basis for the most successful attempts to move 'green' political concerns into the general political process. Having established the grounds for political action based on environmental concerns, it is necessary to consider the form that political action takes, from social movements and non-governmental organisations through to fully independent green political parties.

Environmental politics is not only made by those who are motivated by environmental concern. Even those who generally resist such concerns might find themselves seeking to accommodate them in particular areas. Given the significance of private business to economic growth, its attitude to environmental damage and its response is of great importance. Further, what happens as a result of environmental politics (frequently the product of conflict between those promoting economic development and those promoting environmental concern) – the making of policy – is heavily conditioned by the actions of the bureaucracy and of administrative practice. There are several different ways of making and implementing policy and a variety of ways to evaluate the consequences of administrative responses to environmental problems and policy making.

There is a sense in which comparative study reproduces in analysis the assumptions of national sovereignty. For many political issues this is not important, but for environmental issues it is. Where those issues are either shared or common, where they cross national boundaries or where they are truly global, then what goes on inside the nation-state is not a sufficient focus. Here it is necessary to consider the extent to which international politics and the international system are able to deal with the complexity of global environmental concerns. This necessarily complements the arguments made about understanding the comparative dimension of environmental politics.

Overall, this book has set out the analytical tools that can be used to craft understandings of the sources and character of domestic responses to both local and global environmental problems, as well as the limitations of this response. The same could be said of the international response. The world is confronted by a whole series of environmental problems, especially focused on how to improve the conditions of life through economic growth, while limiting the damage that comes from economic development. In considering the different options that can be applied, the comparative perspective is very useful: it gives a disciplined approach to considering what gives strength or weakness to different systems and different ways of responding to the environmental challenge.

The nature of the political response to environmental challenges has changed over time. Once, political expressions of environmental concern were rare, marginalised and without substantial influence. All that has changed. From concerned individuals a movement seeking change has grown. Green political parties have formed to challenge the ways in which governments respond to the challenge of ecological damage and environmental concern. To a certain extent, environmental issues have moved onto the agenda of normal politics, but new issues arise and new forms of political action come into being to press the claims for environmental concern. Street demonstrations are again taking place, illustrating the depth of concern over the ways in which the rich and powerful have responded to globalisation and its impact on the environment. From Seattle to Melbourne to Prague, people are on the move in disruptive protests, revitalising environmental movements which some thought had declined under the pressure of economic growth and demands for rising standards of living in the affluent world during the 1990s. The internet has become a new means for political mobilisation on environmental issues. So far neither the issues nor the concern have gone away. Environmental politics will not remain static.

Without doubt the most dramatic feature of environmental politics which has changed since the 1990s has been the increased extent of globalisation. Every collective form which we investigated in this book has transnationalised its operations to a large extent, from social movements to NGOs; from political parties to corporations; and even

Plate 15 ***Protest in Manila against transnational mining companies, November 1998. The protests were against these and the Philippines Mining Act of 1995 which provides, amongst other advantages, 5- to 10-year tax breaks to transnational operators. Author's private collection***

institutional and administrative structures are increasingly capable of global responses with real ramifications for nation-states. But, without doubt, the most fascinating phenomenon to come out of this period of rapid internationalisation and globalisation has been one which strikes at the very heart of politics. The very notion of what constitutes the state – both at a national and international level – has been constantly questioned through this exploration of environmental politics.

International environmental governance regimes have recently shifted away from specific multilateral conferences in specific places, such as the Rio Earth Summits. Rather than sovereign states exclusively addressing their environmental problems with some international UN assistance, there are increasingly close relationships between states, huge global environmental NGOs, private companies and global financial institutions such as the World Bank. In the case of Madagascar, for example (mentioned in Chapter 4), such disparate institutions as USAID, the German government, the Japanese government, the French government, the Swiss government, Conservation International, WWF, Wildlife Conservation Society and the World Bank form the international donor consortium which could alternatively be referred to, in effect, as a 'global governance state'. These elite partnerships which intervene in the development and governance of sovereign states will increasingly determine their environmental outcomes. In this picture, sovereignty is not a delimitation of one geographical space over another (nation-states); but is a space 'formed through a series of practices which are defined by an interaction of forces' (Duffy 2006), including some powerful environmental NGOs. In this model, NGOs are just as much part of sovereign, global governance states as national governments, corporations and international administrative systems. This political trend may determine the Earth's future.

Notes

Introduction

1 The new green party experiences in places such as, for example, Taiwan and South Korea are quite fascinating. Their party platforms share with the German Greens the original 'four pillars' of their party constitution: ecology, non-violence and peace, social and environmental justice and equity, and more participative forms of democracy.

2 'North' and 'South' are crude terms that try to name a complex, differentiated and nuanced reality. There are no simple, agreed, unambiguous and effective terms to describe the distribution of power and affluence in the global order. The terms 'North' and 'South' were at their most important back in the days of the Brandt Report and the UNCTAD conferences, where they tried to capture in a geographic/spatial metaphor the maldistribution of power and economic strength between the United States of America, the countries of Western Europe and Japan, in contrast with poorer countries then seeking to industrialise. The spatial metaphor fails as it lumps together the rich and poor within these countries and suggests that they share common perspectives, interests and concerns. Even adding the qualifying 'elites' and 'broad population' does little to make the nomenclature more effective. Parallel terms developed/undeveloped, minority/majority worlds or First World/Third World share similar difficulties. Although we use these terms as a piece of convenient shorthand we are ever mindful of the problems they cause for analysis if they are not treated with extreme caution.

Chapter 1

1 Some of this unreality was picked up by pluralists in their subsequent analytical careers. Charles Lindblom, who was one of the seminal figures in the development of pluralist accounts, went on to write a classic account of the consequences of the preponderance of power held by business in the US political system, *Politics and Markets* (Basic Books, New York, 1977). Dahl himself went on to write a number of works that were concerned with the unequal distribution of economic power and its consequences for a democratic polity; see *After the Revolution: Authority in a Good Society* (Yale University Press, New Haven, 1970) and Economic Democracy (Polity Press, Cambridge, 1985).

2 The concept of 'interests' is one of the most contested in methodological debates of political and social science. There are as many issues to be raised about the conception

and study of interests as there are about the study of power. Some of the elements are covered by Lukes (ch. 6) but see also Connolly (1972) and McEachern (1980).

3 In the vast array of Foucault's work, it is not always clear that there is a unified, consistent 'Foucauldian' interpretation of power, although there are many attempts to state one. The best account is provided by Hindess (1996), who both notes the diversity of Foucault's positions and an important change in his conception of power at about the time he gave his lectures on governmentality and published the *History of Sexuality*, vol. I.

Chapter 2

1 This is not true of the most important of the free market environmentalists, David Pearce, whose work is an exemplary effort at finding market solutions to environmental problems that are treated seriously. Further, David Pearce has sought more than most to take sustainability into the very construction of the basic theorems of neo-classical economics. A most accessible version of the arguments developed by Pearce and his associates is to be found in Cairncross (1991).

2 A joint publication of the International Union for Conservation of Nature and Natural Resources (IUCN), the United Nations Environment Programme (UNEP) and the World Wildlife Fund (now the World-Wide Fund for Nature) (WWF).

3 World Commission on Environment and Development (WCED) chaired by Gro Harlem Brundtland. Australian edition published with additions from the Commission for the Future in 1990.

4 Only some types of eco-feminism can be regarded using this oppositional, paradigmatic model. For example, many parts of eco-feminism are closely related to liberal sensibilities and, as such, are more concerned with seeking eco-feminist change through accommodation within existing political systems (as investigated in the previous section).

5 This section is derived from a more complete treatment of green economics found in Doyle, T. (2001) *Green Power: The Environment Movement in Australia*, University of New South Wales Press, Sydney.

6 Leviathan is a term used most famously by Thomas Hobbes who argued for the necessity of a 'rule of iron' administered by a machine-like monarch. This was the only way in which the earth could avoid catastrophe. Since Hobbes, neo-Hobbsians have emerged advocating similar positions. For a thorough analysis of the neo-Hobbsians read Walker 1988.

7 For an excellent paper of differing approaches by free market environmentalists see Robyn Eckersley's (1993) paper 'Free Market Environmentalism: Friend or Foe?', *Environmental Politics* 2, 1: 1–19.

8 The noted American poet, Gary Snyder, has done much to further the cause of bio-regionalism. For some of his views read Dardick, G. (1986) 'Gary Snyder: Of Place and the Buddha-realm', *Simply Living*, 2, 11: 38–9.

Chapter 3

1 Post-materialism and post-industrialism are not mutually exclusive and share several propositions.

2 This argument about the palimpsest is presented in greater detail in Doyle, T.J. and Kellow, A.J. (1995) *Environmental Politics and Policy Making in Australia*, Macmillan, Melbourne.

3 See Chapters 4 and 8 of Doyle, T.J. and Kellow, A.J. (1995) *Environmental Politics and Policy Making in Australia*, Macmillan, Melbourne, for a close analysis of the politics of informal environmental groups and associations operating in liberal democracies such as Australia.

4 In the United States in the early 1990s, 95 per cent of professional staff and volunteer leaders from over 500 conservation and environment groups nationwide agreed with the following statement: 'Many, perhaps most, minority and poor rural Americans see little in the conservation message that speaks to them.' Not one major environment organisation could boast significant black, Hispanic or Native American membership. See C. Jordan and D. Snow, 'Diversification, Minorities, and the Mainstream Environmental Movement', in Snow, D. (ed.) (1992b) *Voices From the Environmental Movement: Perspectives for a New Era*, Island Press, Washington, DC.

5 As aforesaid, there are numerous exceptions. There are active wilderness-oriented movements, for example, in parts of Africa.

6 In 1990, the United Nations Human Development Report read: 'Poverty is one of the greatest threats to the environment.' In 1993, an International Monetary Fund article read: 'Poverty and the environment are linked in that the poor are more likely to resort to activities that can degrade the environment' (Broad 1994: 812).

7 For an excellent example of this line of reasoning read Hartshom, G.S. (1991) 'Key Environmental Issues for Developing Countries', *Journal of International Affairs*, 3: 393–401.

8 In many well-meaning works prior to the Rio Summit in 1992, it was constantly argued that the South had to become more like the United States in its political system if environmental degradation was to be brought under control. In this vein, the development of 'civil society' is central to the discussion. Underlying this 'civil society' is an unquestioned acceptance of the American form of democracy based on capitalism and global free markets. An excellent example is Ghai, D. and Vivian, J. (eds) (1992) *Grassroots Environmental Action: People's Participation in Sustainable Development*, Routledge, London, New York. They write: 'The existence of a democratic space allowing the expression and defence of community rights and claims has proven to be a crucial factor influencing successful grassroots environmental action . . . The essence of these activities is to persuade or pressure the state to intervene on behalf of the communities through adoption of new legislation' (18–19).

9 In an article by Robin Broad (1994) on the Philippines' environmental movement, 'The Poor and the Environment: Friends or Foe', *World Development*, 22, 6: 812–22, an important distinction is made on the connections between types of poverty and environmental degradation. Broad distinguishes between the 'merely poor' and the 'very, very poor'. Fundamentally, the former category are those still operating subsistance lifestyles (though under threat) and those who have recently been removed from this lifestyle. The latter category are the 'landless and rootless'. These have no security of tenure and little connectedness to place. This category includes those peasants and squatters who survive by cutting forest cover, by consuming wildlife and by planting crops on soils that will erode.

Chapter 4

1 See Carol Deal's booklet prepared for Greenpeace entitled *The Greenpeace Guide to Anti-Environmental Organizations*, Odonian Press, Berkeley, California, 1993. The work looks at six types of 'anti-environmental' organisations, including public relations firms, corporate front groups, think-tanks, legal foundations, endowments, charities, and wise use and share groups. One excellent example of a corporate front group posing as a green NGO is the British Columbia Forest Alliance in Canada. The Canadian timber industry paid Burson-Marstellar one million dollars to create the alliance. Deal writes; 'Like its US counterparts – the Evergreen Foundation and the National Wetlands Coalition – the alliance has two tasks: convincing the public that the current rate of environmental destruction can be maintained or increased without long-term effect, and persuading lawmakers to roll back unprofitable environmental regulations' (16–17).

2 One study performed during glasnost counted 331 environmental organisations in the Russian Federation and the Ukraine (Princen and Finger 1994: 2). Almost no environmental organisations existed in the former Soviet Union before this date. In Indonesia, still playing its politics under an authoritarian regime, environmental NGOs have increased from 79 in 1980 to over 500 in 1992.

3 Earth First! is as non-institutional as organisations get. Some Earth First! participants refer to it as a non-organisation. In fact, it does have many of the attributes of informal groups described in the previous chapter. It falls on the cusp between these two collective forms. But the goals of Earth First! are often listed in its magazine *Earth First!* These goals seem relatively continuous and serve some of the purposes of a constitution.

4 Chatterjee and Finger write of these direct, sometimes militant, eco-actions; 'Yet their tactics have often been surprisingly effective. The Sea Shepherds closed down the Icelandic whaling industry singlehandedly one cold November night in 1986 by the simple expedient of sinking two of its four ships and destroying the refrigeration system of its whale processing plant . . . Ecosaboteurs in Canada blew up a US$4.5 million hydro-electric substation on Vancouver Island in 1982. In Thailand they burnt down a tantalum plant in 1986 causing damage estimated at US$45 million. Lapps in Norway blew up a bridge leading to a dam that had flooded their lands. Then further out from even the deep ecologists are people who do accept physical injury or death as punishment . . . For example, Primea Linea, an Italian group, claimed responsibility for machine gunning Enrio Paoletti, an executive of a Hoffmann-LaRoche subsidiary, who was in charge of the chemical plant in Seveso, Italy, that exploded in 1976 to release a dioxin cloud' (1994: 72).

5 These private purchases of land are common. In 1996, the US Campaign to Save Mount Jumbo in Missoula, Montana, raised sufficient money to buy the elk habitat outright.

6 In 1995, Timothy Doyle interviewed participants enrolled in the United Nations Environment Programme International Certificate in Environmental Management, at the University of Adelaide. The majority of participants worked in the middle and upper management positions in their own country's equivalent of 'environment departments'. All participants were from nations of the 'developing world'. Information derived from this written questionnaire is included in this section. Full results available from the author.

7 For a detailed account of internal politics in Greenpeace, FoE, the ACF and the Wilderness Society see Doyle, T.J. (2000) *Green Power: The Politics of the Environment Movement in Australia*, Scribe Publications, Australia.
8 There are huge regional differences in the operation of FoE. In Australia, for example, there is much autonomy within the organisation, whereas the US experience is far more centralised and hierarchical.
9 See Timothy Doyle (1989), 'Oligarchy in the Conservation Movement: Iron Law or Aluminium Tendency?', *Regional Journal of Social Issues*, summer: 28–47. This article documented an internal power struggle within two environmental organisations in Australia, the ACF and the Wilderness Society, in the late 1980s. Elite theories were drawn upon to provide a theoretical discussion.

Chapter 5

1 There have been minor successes in Eastern Europe. For example, the Green Party in Hungary received 3.7 per cent of the vote at the national level in the elections of spring 1990. This was not enough, however, to secure representation in the parliament (Szabo 1994: 294). Since then 'democratic transition' green parties have formed in Eastern Europe but have not been significant players. Interestingly, several green parties have emerged in the outlying states of the former Soviet Union (Richardson and Rootes 1995: 20).
2 These points were based on empirical studies of the anti-uranium movement and the movement against the flooding of the Franklin River in southwest Tasmania, Australia. Both these movements were extremely active in the late 1970s and 1980s.
3 At the local level, a green councillor, John Gormley, became Lord Mayor of Dublin in early 1994 (Holmes and Kenny 1995: 222).
4 Obviously, there are also many differences between the US and British models. Additionally, within each country there are many different types of election (sometimes working on separate models), ranging from the local level, through the national, to the European Union elections, in the case of Britain; and from local school boards, through the state and federal congresses, to the presidency, in the instance of the USA. The US presidential election is an excellent example of the first-past-the-post system. Each state in the federation has a distinct number of 'electoral college' votes. If one candidate wins the most votes in the state, then he/she takes the entire number of electoral college votes from that state and adds them to his/her national tally. This occurs regardless of the fact that he/she may have the support of only one more elector than another candidate in the state in question. For an introductory guide to this process see O'Connor, K. and Sabato, L.J. (1995) *American Government: Roots and Reform*, second edition, Allyn and Bacon, Boston: 515–27.
5 All states use this system except Tasmania, which utilises the complicated Hare-Clark system. This system is fundamentally one of proportional representation, and it partly accounts for the unusually good showing of Tasmanian greens in this state's electoral polity.
6 We accept that there are contextual differences in the terms 'environmental', 'ecological' and 'green'. But, as mentioned at the outset, for the purposes of establishing a working definition for this book, we use the terms interchangeably.

Chapter 6

1 Gro Harlem Brundtland chaired the UN World Commission on Environment and Development (WCED), which operated from 1983 to 1987 and produced *Our Common Future*.
2 The Rio Earth Summit's formal title was the United Nations Conference on Environment and Development (UNCED). The major communiqué from the conference was *Agenda 21*.

Chapter 7

1 These should be contrasted with David Pearce's unrelated efforts to add a theorem of sustainability to the foundational axioms of neo-classical economics. See Pearce (1991) for the working through of this project and its radical consequences for assessing policy options. In effect, Pearce tries to add a concept of ecological rationality to a field of economic rationality, where rationally linking means and ends is complicated by treating sustainability as if it mattered.
2 For a clear and systematic discussion of this point see Hindess, *Discourses of Power*.
3 The broad shape of these measures has been assessed in Chapter 6, and the comments on those who oppose environmental concern in the name of economic efficiency are not repeated here.
4 Environmentalists from the ACF, the Wilderness Society and Greenpeace initially participated in eight of the nine ESD working groups. Greenpeace withdrew from the process because of proposed Commonwealth resource security legislation. All groups refused to sit on the Forest Industry group from the beginning and pursued their case through RAC instead.
5 C. Hendriks (2002), 'Institutions of Deliberative Democratic Processes and Interest Groups: Roles, Tensions and Incentives', *Australian Journal of Public Administration*, 61, 1: 64–75, 65.
6 D. Stone (1988), *Policy Paradox: The Art of Political Decision Making*, New York and London, W.W. Norton and Co., p. 9.
7 Select Panel of the Public Inquiry into Uranium, 'The Report of the Public Enquiry into Uranium', Adelaide: CCSA, November 1997, pp. 1–40.
8 Although the inquiry was deliberately positioned before a state election, the participants did not directly seek to influence which elected officials would ultimately gain power. Rather, the media circus and interest surrounding the election was seen as an ideal platform from which the silence cloaking the uranium industry could be challenged.

Bibliography

ACF, FoE, GP (2000) 'Dinosaur Business Group is an Embarrassment', media release, 24 May, http://www.geocities.com/jimgreen3/lavoisier.html, accessed 3/09/06.
Adams, G. and Hine, M. (1999) 'Local Environmental Policy Making in Australia', Chapter 10 of *Australian Environmental Policy 2*, pp. 186–203.
Aditjondro, G. and Kowalewski, D. (1994) 'Damning the Dams in Indonesia: A Test of Competing Perspectives', *Asian Survey*, 34, 4: 381–95.
Annis, S. (1992) 'Evolving Connectedness among Environmental Groups and Grassroots Organizations in Protected Areas of Central America', *World Development*, 20, 4: 587–95.
Arendt, H. (1970) *On Violence*, Allen Lane, London.
Australian, The (1996) 6 September.
Australian Business Roundtable on Climate Change (2006) *The Business Case for Early Action*, April.
Bachrach, Peter and Baratz, Morton S. (1962) 'The Two Faces of Power', *American Political Science Review*, 56: 947–52.
—— (1963) 'Decisions and Nondecisions: An Analytical Framework', *American Political Science Review*, 57: 641–51.
—— (1970) *Power and Poverty, Theory and Practice*, Oxford University Press, New York.
Baker, S., Milton, K. and Yearly, S. (eds) (1994) *Protecting the Periphery: Environmental Policy in Peripheral Regions of the European Union*, Frank Cass, Essex.
Bandy, J. and Smith, J. (2005) 'Introduction: Cooperation and Conflict in Transnational Protest,' in J. Bandy and J. Smith (eds), *Coalitions Across Borders: Transnational Protest and the Neoliberal Order*, Rowman and Littlefield, Oxford.
Bansal, P. and Howard, E. (eds) (1997) *Business and the Natural Environment*, Butterworth-Heinemann, Oxford and Boston.
Barnett, J. (2001) *The Meaning of Environmental Security: Environmental Politics and Policy in the New Security Era*, New York, Zed Books.
Barry, J. (1999) *Rethinking Green Politics*, Sage Publications, London.

Barry, J. and Frankland, E.G. (eds) (2002) 'International Environmental Law', *International Encyclopedia of Environmental Politics*, Routledge, London and New York: 283.

Bartlett, Robert V. (1990) 'Ecological Reason in Administration: Environmental Impact Assessment and Administration Theory', in Robert Paehlke and Douglas Torgerson (eds), *Managing Leviathan: Environmental Politics and the Administrative State*, Belhaven Press, London: 81–96.

Baviskar, A. (1995) *In the Belly of the River: Tribal Conflict over Development in the Narmada Valley, Delhi*, Oxford University Press, New York.

Bean, C. and Kelley, J. (1995) 'The Electoral Impact of New Politics Issues: The Environment in the 1990 Australian Federal Election', *Comparative Politics*, April: 339–56.

Bebbington, A. and Thiele, G. with Davies, P., Prager, M. and Riveros, H. (1993) *Non-Governmental Organizations and the State in Latin America: Rethinking Roles in Sustainable Agricultural Development*, Routledge, London and New York.

Beckerman, Wilfred (1974) *In Defence of Economic Growth*, Jonathan Cape, London.

—— (1990) *Pricing For Pollution* (2nd edn), Institute for Economic Affairs, London.

—— (1995) *Small is Stupid: Blowing the Whistle on the Greens*, Duckworth, London.

Bell, S. (1986) 'Socialism and Ecology: Will Ever the Twain Meet', *Social Alternatives*, 6, 3: 5–12.

Benedict, R. (1991) *Ozone Diplomacy: New Directions in Safeguarding the Planet*, Harvard University Press, Cambridge, MA.

Bennett, J. and Block, W. (1991) *Reconciling Economics and the Environment*, Australian Institute for Public Policy, West Perth.

Bernstorff, A. and Stairs, K. (2001) *POPs in Africa: Hazardous Waste Trade 1980–2000*, Greenpeace International, Amsterdam.

Bettinga, J. and Hollinger, R. (2001) 'Meet Your Greens', *New Statesman*, 130, 4546: 34.

Blair, A. and Hitchcock, D. (2001) *Environment and Business*, Routledge, London and New York.

Boggs, C. (1986) *Social Movements and Political Power: Emerging Forms of Radicalism in the West*, Temple University Press, Philadelphia.

Bookchin, M. (1980) 'Ecology and Revolutionary Thought', in R. Roelofs, J. Crowley and D. Hardest (eds), *Environment and Society*, Prentice-Hall, New Jersey.

—— (1987) 'Social Ecology versus "Deep Ecology": A Challenge for the Ecology Movement', *The Raven Anarchist Quarterly*, 1, 3: 219–50.

Bookchin, M. and Foreman D. (1991) *Defending the Earth: A Dialogue between Murray Bookchin and Dave Foreman*, South End Press, Boston.

Boreham, G. (1996) 'Keating Woos Greens', *The Age*, 25 January.

Bramwell, A. (1994) *The Fading of the Greens: The Decline of Environmental Politics in the West*, Yale University Press, New Haven, CT, and London.
Broad, R. (1994) 'The Poor and the Environment: Friends or Foes?', *World Development*, 22, 6: 812–22.
Brodine, V.W. (1992) 'Green Cuba', *Multinational Monitor*, 13, 11: 23–5.
Brown, L.D and Fox, J. (1998) 'Accountability within Transnational Coalitions', in J.A. Fox and L.D. Brown (eds), *The Struggle for Accountability: The World Bank, NGOs, and Grassroots Movements*, MIT Press, Cambridge, MA and London: 439–83.
Bryant, R.L. and Bailey, S. (1997) *Third World Political Ecology*, Routledge, London.
Bullard, R. (1990) *Dumping in Dixie*, Boulder, CO: Westview Press.
—— (ed.) (1993) *Confronting Environmental Racism: Voices from the Grassroots*, South End Press, Boston.
Burchell, G., Gordon, C. and Miller P. (eds) (1991) *The Foucault Effect: Studies in Governmentality*, Harvester Wheatsheaf, London.
Bureau of Industry Economics (1992) *Environmental Regulation: The Economics of Tradeable Permits – A Survey of Theory and Practice*, Research Report 42, AGPS, Canberra.
Buttall, F.H. (2005) 'The Environmental and Post-Environmental Politics of Genetically Modified Crops and Foods,' *Environmental Politics*, 14, 3: 309–23.
Cairncross, Francis (1991) *Costing the Earth*, Economist Books, London.
Callenbach, E. *et al.* (1993) *EcoManagement: The Elmwood Guide to Ecological Auditing and Sustainable Business*, Berrett-Koehler, San Francisco.
Calvert, P. and Calvert, S. (1999) *The South, the North and the Environment*, Pinter, London and New York.
Capling, A. and Galligan, B. (1991) 'Beyond the Protective State: The Political Economy of Australians: Manufacturing Industry Policy', *Australian and New Zealand Journal of Sociology*, 119, 30, 2: 220–1.
Carlassare, E. (1994) 'Essentialism in Ecofeminist Discourse', in C. Merchant (ed.), *Ecology: Key Concepts in Critical Theory*, Humanities Press, New Jersey.
Carothers, T. (2002) 'The End of the Transition Paradigm', *Journal of Democracy*, 13, 1: 5–21.
Carruthers, J. (1995) *The Kruger National Park: A Social and Political History*, University of Natal Press, Pietermaritzburg.
Carter, F.W. and Turnock, D. (1993) *Environmental Problems in Eastern Europe*, Routledge, New York.
Catchlove, J. (2006) *Seeking Sustainable Solutions to Climate Change*, Friends of the Earth Adelaide, Adelaide.
Chaloupka, B. (1996) 'The Year of the Green', *Missoula Independent*, 28 March: 17.
Chapin, M. (2004) 'A Challenge to Conservationists', *World Watch*, November–December, World Watch Institute, 17–31.

Chatterjee, P. and Finger, M. (1994) *The Earth Brokers: Power, Politics and World Development*, Routledge, London and New York.

Cheney, J. (1989) 'Postmodern Environmental Ethics: Ethics as Bioregional Narrative', *Environmental Ethics*, 11: 117–34.

Chiapaslink (2000) *The Zapatistas: A Rough Guide*, Chiapaslink and Earthright Publications, Bristol.

Clark, M. and Netherwood, A. (1999) 'Beyond Volunteering and Rhetoric: Implications of Local Agenda 21 initiatives in Wales', in S. Buckingham-Hatfield and S. Percy (eds), *Constructing Local Environmental Agendas: People, Places + Participation*, Routledge, London and New York: 42–55.

Cock, J. and Koch, E. (eds) (1991) *Going Green: People, Politics and the Environment in South Africa*, Oxford University Press, Cape Town.

Conley, V.A. (1997) *Ecopolitics: The Environment in Poststructuralist Thought*, Routledge, London and New York.

Connelly, J. and G. Smith (1999) *Politics and the Environment: From Theory to Practice*, Routledge, London.

Connolly, W.E. (1972) 'On "Interests" in Politics', *Politics and Society*, 2: 459–77.

Coonan, C. (2006) 'Three Gorges Dam activist who defied state is paralysed in attack', *The Independent (London)*, 14 June, http://threegorgesprobe.org/tgp/index.cfm?DSP=content &ContentID=15648

Cooper, M. (1994) 'The Greens Climb in New Mexico', *The Nation*, 259, 3: 453–7.

Coyote, H. (1991) 'The Corporate Takeover of Friends of the Earth', *Chain Reaction*, 63/64, April: 35–8.

Crenson, M.A. (1971) *The Un-Politics of Air Pollution: A Study of Non-Decision Making in the Cities*, Johns Hopkins University Press, Baltimore.

Crick, B. (1964) *In Defence of Politics*, Allen Lane, London.

Cuomo, C. (1992) 'Unravelling the Problems in Ecofeminism', *Environmental Ethics*, winter: 351–63.

Curtis, B. (1995) 'Taking the State Back Out: Rose and Miller on Political Power', *British Journal of Sociology*, 46, 4: 575–89.

Curtis, J. (1998) 'The Challenge of Environmental Security', *Habitat Australia*, 26, 6: 32.

Cushman, J.H. (1996) 'Adversaries Back the Current Rules Curbing Pollution', *New York Times*, Monday, 12 February, pp. 1 and C11.

Dabelko, G.D. (1999) 'The Environmental Factor', *The Wilson Quarterly*, 23, 4: 14.

Dabelko, G.D. and Dabelko, D.D. (1995) 'Environmental Security: Issues of Conflict and Redefinition', *Woodrow Wilson Environmental Change and Security Project Report*, 1: 3–12.

Dahl, R.A. (1957) 'The Concept of Power', *Behavioural Science*, 2, 3: 201–15.

—— (1961) *Who Governs? Democracy and Power in an American City*, Yale University Press, New Haven, CT, and London.

—— (1970) *Modern Political Analysis*, Prentice-Hall, New Jersey.

Dalby, S. (2002) 'Questioning Geopolitics: Political Projects in a Changing World System', *Contemporary Sociology*, 31, 3: 316–17.

Dalton, R.J. (1994) *The Green Rainbow: Environmental Groups in Western Europe*, Yale University Press, New Haven, CT, and London.

Darier, E. (ed.) (1999) *Discourses of the Environment*, Blackwell, Oxford.

Dawdy, P. (2000) 'Spotted Owls on the Outs: Supposedly Saved, Biologists Now Count Half as Many on Federal Land', *Seattle Weekly*, 4 September, http://www.seattleweekly.com/ news/0236/news-dawdy.php, accessed 10/09/06.

De Shalit, A. and Talias, M. (1994) 'Green or Blue or White? Environmental Controversies in Israel', *Environmental Politics*, 3, 2: 273–94.

Deal, C. (1993) *The Greenpeace Guide to Anti-Environmental Organizations*, Odonian Press, Berkeley, CA.

Dean, M. (1991) *The Constitution of Poverty*, Routledge, London.

Department of Home Affairs and Environment (1982) *A National Conservation Strategy for Australia*, AGPS, Canberra.

Derbyshire, I. (1987) *Politics in West Germany: From Schmidt to Kohl*, W.R. Chambers, London.

Detlef, J. (1994) 'Unifying the Greens in a United Germany', *Environmental Politics*, 3, 2: 312–18.

Devall, B. and Sessions, G. (1985) *Deep Ecology*, Peregrine Smith Books, Salt Lake City, UT.

Diamond, L. (2002) 'Elections without Democracy: Thinking about Hybrid Regimes', *Journal of Democracy*, 13, 2: 21–35.

Diani, M. (1992) 'The Concept of Social Movement', *Sociological Review*, 40, 1.

Dienel, P. and Renn, O. (1995) 'Planning Cells: A Gate to Practical Mediation', in O. Renn, T. Webber and P. Wiederman, *Fairness and Competence in Citizen Participation*, Klewer Academic, Boston.

DiLorenzo, Thomas J. (1993) 'The Mirage of Sustainable Development', *The Futurist*, September–October: 14–19.

Ditz, D., Ranganathan, J. and Banks, D. (1995) *Green Ledgers: Case Studies in Corporate Environmental Accounting*, World Resources Institute, Washington, DC.

Dobson, A. (1995) *Green Political Thought* (2nd edn), Routledge, London and New York.

—— (2004) *Citizenship and the Environment*, Oxford University Press, Oxford.

—— (2006) 'Ecological citizenship: A Defence', *Environmental Politics*, 15, 3: 447–51.

Dodds, F. and Pippard, T. (2005) *Human and Environmental Security: An Agenda for Change*, Sterling VA/Earthscan, London.

Doherty, B. (1992) 'The Fundi-Realo Controversy: An Analysis of Four European Green Parties', *Environmental Politics*, 1, 1: 95–120.

—— (2002) *Ideas and Actions in the Green Movement*, Routledge, London.

Doherty, B. (2006) 'Friends of the Earth International: Negotiating a Transnational Identity', *Environmental Politics*, special guest issue edited by B. Doherty and T. Doyle, 'Beyond Borders: Environmental Movements and Transnational Politics', 15, 5: 860–80, Taylor and Francis, London.

Doherty, B. and Doyle, T. (2006) 'Environmental Movements', *Environmental Politics*, special issue guest edited by B. Doherty and T. Doyle, 15, 4, Taylor and Francis, London.

Doherty, B., Paterson, M. and Seel, B. (2000) *Direct Action in British Environmentalism*, Routledge, London.

Donati, P. (1984) 'Organization between Movement and Institution', *Social Science Information*, Sage, London, Beverly Hills and New Delhi, 23, 4/5: 837–59.

Douglas, M., Lee, Y.S.F. and Lowry, K. (1994), 'Introduction to the Special Issue on Community Based Urban Environmental Management in Asia', *Asian Journal of Environmental Management*, 2, 1: ix–xv.

Dowdeswell, E. (1994) Speaking Notes for Elizabeth Dowdswell, Under-Secretary General and Executive Director, United Nations Environment Programme, electronic preparation by the Population Information Network of the United Nations Population Division in collaboration with the United Nations Development Programme, 6 September.

Dowie, M. (1994) 'The Selling (Out) of the Greens', *The Nation*, 18 April: 514–18.

—— (1995a) *Losing Ground: American Environmentalism at the Close of the Twentieth Century*, MIT Press, Cambridge, MA, and London.

—— (1995b) 'The Fourth Wave: An Opinion by Mark Dowie', *Mother Jones*, March–April: 34–6.

Doyle, T. (1986) 'The "Structure" of the Conservation Movement in Queensland', *Social Alternatives*, 5, 2: 27–32.

—— (1989) 'The Conservation Movement and the Aluminium Tendency of Oligarchy', *Regional Journal of Social Issues*, summer: 28–47.

—— (1991) 'Informal Groups and the Conservation Movement: A Matter of Introspection', *Conference of the Australian Political Studies Association*, Griffith University, 17–19 July.

—— (1994a) 'Direct Action in Environmental Conflict in Australia: A Re-examination of Non-Violent Action', *Regional Journal of Social Issues*, Australia, 28: 1–13.

—— (1994b) 'Dissent within the Environment Movement', *Social Alternatives*, 13, 2: 24–6.

—— (1996) 'Agenda 21: Ecological Imperialism and the Globalisation of Environmental Management', *Ecopolitics 10*, Proceedings of the Conference, Australian National University, September.

—— (1998) 'Sustainable Development and Agenda 21: The Secular Bible of Global Free Markets and Pluralist Democracy', *Third World Quarterly*, 19, 4: 771–86.

—— (1999) 'Resource Decision-making in Arid Lands: The Consequences of Wise Use', in K. Walker and K. Crowley (eds), *Environmental Policy Two*, University of New South Wales Press, Sydney.

—— (2000) *Green Power: The Environment Movement in Australia*, University of New South Wales Press, Sydney.

—— (2004). 'Dam Disputes in Australia and India: Appreciating Differences in Struggles for Sustainable Development', in D. Gopal and D. Rumley (eds), *India and Australia: Issues and Opportunities*, Authorspress, New Delhi.

—— (2005) *Environmental Movements in Majority and Minority Worlds: A Global Perspective*, Rutgers University Press, New Brunswick, New Jersey and London.

—— (2005b) 'Outside the State: Australian Green Politics and the Public Inquiry into Uranium', in R. Paehlke and D. Torgerson (eds), *Managing Leviathan: Environmental Politics and the Administrative State*, Broadview Press, Peterborough, Canada: 235–56.

Doyle, T. and Kellow, A. (1995) *Environmental Politics and Policy-making in Australia*, Macmillan, Melbourne.

Doyle, T. and Walker, K. (1996) 'Looking for a World They Can Call Their Own', *Campus Review*, Australia, 1–7 February: 10, 16.

Doyle, T. and Doherty, B.J. (2006) 'Green Public Spheres and the Green Governance State: The Politics of Emancipation and Ecological Conditionality', *Environmental Politics*, special Issue guest edited by B. Doherty and T. Doyle, 15, 4, Taylor and Francis, London.

Doyle, T. and Simpson, A.J. (2006) 'Traversing More than Speed Bumps: Green Politics under Authoritarian Regimes in Burma and Iran', *Environmental Politics*, special issue guest edited by B. Doherty and T. Doyle, 15, 4, Taylor and Francis, London.

Doyle, T. and Risely, M. (eds) (2008) *Crucible for Survival: Environmental Justice and Security in the Indian Ocean Region*, Rutgers University Press, New Jersey and London.

Dryzek, J. (1987) *Rational Ecology: Environment and Political Economy*, Basil Blackwell, Oxford.

—— (1990) 'Design for Environmental Discourse: The Greening of the Administrative State?', in R. Paehlke and D. Torgerson (eds), *Managing Leviathan: Environmental Politics and the Administrative State*, Belhaven Press, London: 97–111.

—— (1992) 'The Good Society versus the State: Freedom and Necessity in Political Innovation', *Journal of Politics*, 54, 2: 518.

—— (1997, 2005) *The Politics of the Earth: Environmental Discourses*, Oxford University Press, Oxford.

Dryzek, J. and Torgerson, D. (1992) 'Democracy and the Policy Sciences: A Progress Report,' editorial, *Policy Sciences*, 26: 127–37.

Duffy (2006) 'Madagascar . . .' *Beyond Borders*, *Environmental Politics*, 15, 4.

Duverger, M. (1972) *The Study of Politics*, Crowell, New York.

Earth Charter Drafting Team (1999), 'Earth Charter General Principles Draft 1999', Canberra, February.

Earth Times Foundation (1996) *The Earth Times*, 9, 3, Geneva and New York.

Easterbrook, G. (1995) *A Moment on the Earth: The Coming of Age of Environmental Optimism*, Viking Penguin, New York.

Ebrahim, A. (2003) 'Accountability in Practice: Mechanisms for NGOs', *World Development*, 31, 5: 813–29.

Eckersley, R. (1990) 'The Ecocentric Perspective', in C. Prybus and R. Flanagan (eds), *The Rest of the World is Watching*, Pan Macmillan, Sydney.

—— (1992) *Environmentalism and Political Theory: Toward an Ecocentric Approach*, UCL Press, London.

EcoGeneration (2006) 'The Asia Pacific Partnership on Clean Development and Climate: What It Is and What It Isn't', *EcoGeneration*, February–March: 6–9.

Ecologically Sustainable Development Working Groups (1991) *Final Reports*, AGPS, Canberra.

Ecologically Sustainable Development Working Groups Chairs (1992) *Greenhouse Report*, AGPS, Canberra.

Economist, The (1994) 10 September, 332: 51–2.

—— (1995) 10 June, 335: 46.

Economy, E. (2004) *The River Runs Black:The Environmental Challenge in China's Future*, Ithaca, Cornell University Press.

Edelman, M. (1964) *The Symbolic Uses of Politics*, University of Illinois Press, Chicago.

Ekins, P. (1992) *A New World Order: Grassroots Movements for Global Change*, Routledge, London.

Elliott, L. (1998) *The Global Politics of the Environment*, Macmillan, Basingstoke.

Environmental News Network (2006) '"Carbon dioxide . . . we call it life", US TV ads say', 18 May, http://www.enn.com/today.html?id=10481, accessed 12/09/06.

EPA (2006), 'Clear Skies Act of 2003', Environmental Protection Authority website, http://www.epa.gov/air/clearskies/fact2003.html, accessed 19/09/06.

ERA (2005) *The Shell Report: Continuing Abuses in Nigeria – Ten Years After Ken Saro-Wiwa*, FoENigeria, Benin City, Nigeria.

Esteva, G. (2003) 'A Flower in the Hands of the People', *New Internationalist*, 360, September.

Eureka Times-Standard (2004) '9th Circuit Ruling Buttresses Spotted Owl Protection', *Eureka-Times Standard*, 7 August, http://www.times-standard.com/Stories/0,1413,127~ 2896~2320647,00.html 7aug04, accessed 10/09/06.

Faehmann quoted in Wikinews, 12 Jan. 2006.

Fagan, A. (1994) 'Environmentalism and Transition in the Czech Republic', *Environmental Politics*, fall, 3: 479–94.

—— (2006) 'Neither "North" Nor "South": The Environment and Civil Society in Post-conflict Bosnia–Herzegovina', *Environmental Politics*, 15, 5.

Ferry, L. (1992) *The New Ecological Order*, University of Chicago Press, Chicago and London.

Fig, D. (1991) 'Flowers in the Desert: Community Struggles in Namaqualand', in J. Cock and E. Koch (eds), *Going Green: People, Politics and the Environment in South Africa*, Oxford University Press, Cape Town.

Fischer, K. and Schot, J. (1993) *Environmental Strategies for Industry: International Perspectives on Research Needs and Policy Implications*, Island Press, Washington, DC.

Fitch, J.E. (2002) 'Spotted Owl Controversy', in J. Barry and E.G. Frankland (eds), *International Encyclopedia of Environmental Politics*, Routledge, London and New York: 432.

Flannery, T.F. (1994) *The Future Eaters: Ecological History of the Australasian Lands and People*, Reed Books, Kew, Victoria.

Foreman, D. (n.d.) 'A Spanner in the Woods', interviewed by B. Devall, *Simply Living*, 2, 12: 40–3.

Foucault, M. (1984) *History of Sexuality*, vol. I, Peregrine Books, London.

—— (1991) 'Governmentality', in Graham Burchell, Colin Gordon and Peter Miller (eds), *The Foucault Effect: Studies in Governmentality*, Harvester Wheatsheaf, London: Chapter 4.

Fox, W. (1990) *Toward a Transpersonal Ecology: Developing New Foundations for Environmentalism*, Shambala, Boston.

Frankel, C. (1998) *In Earth's Company: Business, Environment and the Challenge of Sustainability*, New Society Publishers, Gabriola Is., Canada.

Frankland, E.G. (1995) 'Germany: The Rise and Fall and Recovery of Die Grünen', in D. Richardson and C. Rootes (eds), *The Green Challenge*, Routledge, London and New York.

Friends of the Earth International and BUND (2002) 'Earth Summit: Betrayal', *From Rio to Jo'burg: what's the news?* website, http://www.rio-plus-10.org/, accessed 10/09/06.

Gane M. and Johnson, T. (eds) (1993) *Foucault's New Domains*, Routledge, London: 75–105.

Gare, A.E. (1995) *Postmodernism and the Environmental Crisis*, Routledge, London and New York.

Gaventa, J. (1980) *Power and Powerlessness: Quiescence and Rebellion in an Appalachian Valley*, Clarendon Press: Oxford.

Gedicks, A. (1993) *The New Resource Wars: Native and Environmental Struggles against Multinational Corporations*, South End Press, Boston.

Ghai, D. and Vivian, J. (eds) (1992) *Grassroots and Environmental Action: People's Participation in Sustainable Development*, Routledge, London and New York.

Global Environmental Facility Black Sea Environmental Programme (1994) International Black Sea NGO Forum Meeting, *Report*, 7–10 November.

—— (1995a) *International Black Sea NGO Forum Conclusions and Recommendations*, 16–18 October.

—— Programme Coordination Unit (1995b) *Commercial Fisheries in the Black Sea: Three Decades of Decline*, leaflet.

—— (1995c) *Black Sea NGO Directory, Directory of Environmental NGOs in Bulgaria, Georgia, Romania, Russia, Turkey and Ukraine.*

—— (1995d) *Saving the Black Sea*, official newsletter, October, issue 3.

Global Greens (2006), *Global Greens* website, http://www.globalgreens.info/, accessed 17/05/06.

Goldberg, K. (1994) 'Green Relief for Forest Defenders', *The Progressive*, 58, 3: 13.

Golding, W. (1954) *Lord of the Flies: A Novel*, London.

Goodwin, B. (1992) *Green Political Theory*, Polity Press, Cambridge.

Gore, A. (1992) *Earth in Balance: Ecology and the Human Spirit*, Houghton Mifflin, Boston.

Green Party of England & Wales (2002) 'Greens Lead in Anti-Euro Campaign', *Synthesis/Regeneration 29: A Magazine of Green Social Thought*, fall: 45–6.

Green Party Taiwan (2006) 'History of Green Party Taiwan', Green Party Taiwan website, http://www.greenparty.org.tw/english.php?itemid=104&catid=21#more, accessed 30/08/06.

Greenpeace (1998) 'Japanese Plans to Make Caribbean the Toxic Throughway for Clandestine Shipments of Nuclear Waste and Plutonium', unpub. briefing, Adelaide, April.

Griggs, S. and Howarth, D. (1999) 'New Social Movements and the Politics of Environmental Protest: The Campaign against Manchester Airport's Second Runway', paper presented to the 1999 ECPR Joint Sessions of Workshops, March, Mannheim, Germany.

Hamilton, C. (2002) 'Green Conspiracy Theory', *Canberra Times*, 10 January.

Hansen, C. (1995) 'Sierra Club Management Shaming Muir's Memory', *Earth First!*, Eostar edition: 26.

Hardin, G. (1968) 'The Tragedy of the Commons', *Science*, 162: 1,243–8.

Hartshorn, G.S. (1991) 'Key Environmental Issues for Developing Countries', *Journal of International Affairs*, 3: 393–402.

Hawken, P. (1993) *The Ecology of Commerce: A Declaration of Environmental Sustainability*, Harper Business, New York.

Hawken, P. Lovins, A. and Lovins, L.H. (1999) *Natural Capitalism: Creating the Next Industrial Revolution*, Little, Brown & Co, Boston.

Hay, P. (2004) *Main Currents in Western Environmental Thought*, University of New South Wales Press, Sydney.

Hay, P. and Haward, M. (1988) 'Comparative Green Politics: Beyond the European Context?', *Political Studies*, 36: 433–48.

Hayes, G. (1994) 'In Splendid Isolation: The French Greens', *Environmental Politics*, 3: 169–73.

Healthy Forests (2006), home page, Healthy Forests website, http://www.healthyforests.gov/, accessed 19/09/06.

Helman, Z. (2006) 'Opposition to GMOs: A Comparative Case Study of Social

Movements in Germany and India', postgraduate paper presented in the School of History and Politics at the University of Adelaide.
Helvarg, D. (1994) *The War Against the Greens: The 'Wise Use' Movement, the New Right, and Anti-Environmental Violence*, Sierra Club Books, San Francisco.
Hendriks, C. (2002) 'Institutions of Deliberative Democratic Processes and Interest Groups: Roles, Tensions, and Incentives', *Australian Journal of Public Administration*, 61, 1.
Herring, R.J. (2005) 'Miracle Seeds, Suicide Seeds, and the Poor: GMOs, NGOs, Farmers and the State', in R. Ray and M.F. Katzenstein (eds), *Social Movements in India: Poverty, Power and Politics*, Oxford University Press, New Delhi: 203–32.
Heschel, S. (1995) 'Feminists Gain at Population Conference', *Dissent*, winter: 15.
Hindess, B. (1996) *Discourses of Power: From Hobbes to Foucault*, Basil Blackwell. Oxford.
Hirsch P. and Warren C. (1998) *The Politics of South East Asia: Resources and Resistance*, Routledge, London.
Hirsch, P. and Lohmann, L. (1989) 'Contemporary Politics of Environment in Thailand', *Asian Survey*, 29, 4: 439–51.
Ho, M.-W., Ching, L.L. *et al.* (eds) (2003) *The Case for a GM-Free Sustainable World*, Independent Science Panel, London, Institute of Science in Society, Penang, Third World Network.
Ho, M.-S. (2003), 'The Politics of Anti-nuclear Protest in Taiwan: A Case of Party-Dependent Movement (1980–2000)', *Modern Asian Studies*, 37, 3: 683–708.
Holmes, R. and Kenny, M. (1995) 'The Electoral Breakthrough of the Irish Greens?', *Environmental Politics*, 3: 218–26.
Homer-Dixon, T. (1994) 'Population and Conflict', distinguished lecture series on population and development, International Union for the Scientific Study of Population, Belgium.
Hornborg, A. (1994) 'Environmentalism, Ethnicity and Sacred Places: Reflection on Modernity, Discourse and Power', *Canadian Review of Sociology and Anthropology*, 31, 008–4948: 245–67.
Howes, M. (2005) *Politics and the Environment: Risk and the Role of Government and Industry*, Allen and Unwin.
Hunter, B., Siegfried, R. and Sunter, C. (1989) *South African Environments into the 21st Century*, Human & Rousseau and Tafelberg, Cape Town.
Hutchings, V. (1994a) 'Green Gauge: Racism Plays a Key Role in Who Gets Toxic Waste Dumped on Them', *New Statesman and Society*, 11 March: 31.
—— (1994b) 'Support Your Local Village Green', *New Statesman and Society*, 11 March: 20–2.
Hyndman, D. (1991) 'Zipping Down the Fly on the OK Tedi Project', in J. Connell and R. Howitt (eds), *Mining and Indigenous Peoples in Australasia*, Sydney University Press, Sydney: Chapter 12.

Imura, H. (1994) 'Japan's Environmental Balancing Act', *Asian Survey*, 34, 4: 355–68.
Industry Commission (Australia) (1991) *Costs and Benefits of reducing Greenhouse Gas Emissions*, AGPS, Canberra.
Inglehart, R. (1977) *The Silent Revolution*, Princeton University Press, New Jersey.
—— (1990) *Culture Shift in Advanced Industrial Society*, Princeton University Press, New Jersey.
International Union for Conservation of Nature and Natural Resources [IUCN], United Nations Environment Programme [UNEP] and the World Wildlife Fund [WWF] (1980) *World Conservation Strategy: Living Resource Conservation for Sustainable Development*, Morges, Switzerland.
Ip, Po-keung (1983) 'Taoism and the Foundations of Environmental Ethics', *Environmental Ethics*, 5, 4: 335–43.
Jaensch, D. (1983) *The Australian Party System*, George Allen & Unwin, Sydney.
Jahn, D. (1994) 'Unifying the Greens in a United Germany', *Environmental Politics*, 3, 2.
Jesinghausen, M. (1995) 'General Election to the German Bundestag on 16 October 1994: Green Pragmatists in Conservative Embrace or a New Era for Green Party Democracy', *Environmental Politics*, 4, 1: 108–14.
Johnson, S. (1995) *The Politics of Population: The International Conference on Population and Development Cairo 1994*, Earthscan Publications, London.
Johnston, H., Larana, E. and Gusfield, J. (eds) (1995) 'New Social Movements: From Ideology to Identity', *Social Forces*, 73, 4: 1633–5.
Jones, T. (2004) 'Campbell, Thompson Debate Tas Forest Policies', *Lateline* TV programme transcript, Australian Broadcasting Corporation, 7 October.
Joppke, C. and Markovits, A. (1994) 'Green Politics in the New Germany: The Future of an Anti-Party', *Dissent*, spring: 235–40.
Jordan, A. (1994) 'Reviewing Rio: A Plummet from the Summit?', *Environmental Politics*, spring: 159–63.
Kagin, M. (1995) 'Alternative Politics: Is a Third Party the Way Out', *Dissent*, winter: 22–6.
Kalland, A. and Persoon, G. (eds) (1998) *Environment Movements in Asia*, Curzon, Surrey.
Karen Human Rights Group (1996) *Effects of the Gas Pipeline Project*, 23 May, http:// www.ibiblio.org/freeburma/humanrights/khrg/archive/khrg96/khrg9621.html, accessed 18/04/06.
Keck, M.E and Sikkink, K. (1998) *Activists Beyond Borders: Advocacy Networks in International Politics*, Cornell University Press, Ithaca and London.
Kerenyi, S. and Szabó, M. (2006) 'Transnational Influences on Patterns of Mobilization within Environmental Movements in Hungary', *Environmental Politics*. 15, 5.

Khan, F. (1994) 'Rewriting South Africa's Conservation History: The Role of the Native Farmers Association', *Journal of South African Studies*, 20, 4.
KHRG (1996). *Effects of the Gas Pipeline Project*, Karen Human Rights Group, 23 May, http://www.ibiblio.org/freeburma/humanrights/khrg/archive/khrg96/khrg9621.html, accessed 18/04/06.
Kilby, P. (2006) 'Accountability for Empowerment: Dilemmas Facing Non-Governmental Organisations', *World Development*, 34, 6: 951–63.
Kingsnorth, P. (2003a) 'The Global backlash,' *New Statesman*, 132, 4635: 18–20.
—— (2003b) *One No, Many Yeses: A Journey to the Heart of the Global Resistance Movement*, Free Press, UK.
Klandermans, B. and Tarrow, S. (1988) 'Mobilization into Social Movements: Synthesizing European and American Approaches', in B. Klandermans, H. Kriesi and S. Tarrow (eds), *From Structure to Action, Comparing Social Movement Research Across Cultures*, JAI Press, Greenwich.
Klein, N. (2002) *Fences and Windows: Dispatches from the Front Lines of the Globalization Debate*, Flamingo, London: 208–23.
Knup, E. (1997) *Environmental NGOs in China: A Overview*. Woodrow Wilson Centre Press, Washington, DC.
Kraft, M.E. (1995) 'Legislation: United States', in R. Paehlke (ed.), *Conservation and Environmentalism: An Encyclopedia*, Fitzroy Dearborn, London and Chicago: 409–12.
Kramer, D. (1994) 'The Graying of the German Greens: Environmental Politics and the Crisis of Consensus', *Dissent*, spring: 231–4.
Krummer, K. (1999) *International Management of Hazardous Wastes*, Oxford University Press, Oxford.
Kuhn, T.S. (1969) *The Structure of Scientific Revolutions*, University of Chicago Press, Chicago.
Kütting, G. (2000) *Environment, Society and International Relations: towards more effective international environmental agreements*, Routledge, London and New York.
Lady, M. (1994) 'Greening Party Politics in the US and Cuba', paper presented in School of Social Sciences, Discipline of Geography and Environmental Studies, University of Adelaide.
Larana, E., Johnston, H. and Gusfield, J.R. (eds) (1992) *New Social Movements: From Ideology to Identity*, Temple University Press, Philadelphia.
Lavoisier Group (2006) website http://www.lavoisier.com.au, accessed 3/09/06.
Leftwich, A. (1983) *Redefining Politics: People, Resources and Power*, Methuen, London.
Leip, D. (2005), *Atlas of US Presidential Elections* website, http://uselectionatlas. org/, accessed 4/09/06.
Lester, J.P. (1998) 'Looking Backward to See Ahead', *Forum for Applied Research & Public Policy*, 13, 4, 30.
LETS South Australia (1997) Pamphlet, Local Exchange Trading Systems (LETS), Adelaide.

Lewis, M.W. (1992) *Green Delusions: An Environmentalist Critique of Radical Environmentalism*, Duke University Press, Durham, NC.
—— (1996) 'Radical Environmental Philosophy and the Assault on Reason', in P. Gross, N. Levitt and M.W. Lewis (eds), *The Flight from Science and Reason, Annals of the New York Academy of Science*, 775.
Lindblom, Charles (1977) *Politics and Markets*, Basic Books, New York.
List, P.C. (1993) *Radical Environmentalism: Philosophy and Tactics*, Wadsworth Publishing Company, Belmont, CA.
Lomborg, B. (2001) *The Skeptical Environmentalist: Measuring the Real State of the World*, Cambridge University Press, Cambridge.
Lowe, E. (1996) 'Industrial Ecology: A Context for Design and Decision', in J. Fiskel (ed.), *Design for Environment*, McGraw-Hill, New York.
Lowe, P. and Goyder, J. (1983) *Environmental Groups in Politics*, Allen & Unwin, London and Boston.
Lukes, S. (1974) *Power: A Radical View*, Macmillan, London.
Lundstedt, H. (2004) 'The Green Party and NGOs: The Environment as a Political Issue in Sweden', paper presented in School of Social Sciences, Discipline of Geography and Environmental Studies, University of Adelaide.
Ma'anit, A. (2006) 'CO_2NNED Keynote: If You Go Down to the Woods Today . . .', *New Internationalist*, 391, July: 2–6.
MacAndrews, C. (1994) 'Politics of the Environment in Indonesia', *Asian Survey*, 34, 4: 369–80.
Maddox, J. (1972) *The Doomsday Syndrome*, Macmillan, London.
Malthus, T.H. (1826) *An Essay on the Principles of Population* (6th edn), Murray, London.
Martin, B. (1984) 'Environmentalism and Electoralism', *The Ecologist*, 14, 3: 110–18.
Martinez-Alier, J. (2004) 'Ecological Distribution Conflict and Indicators of Sustainability', English translation in *Revista iberio-americana de economía ecológica*, 1, 21–31.
Marx, K. (1967) *Capital*, 3 vols, International Publishers, New York.
Marx, K. and Engels, F. [1848] (1972) *Kommunistische Manifest* [*The Communist Manifesto*] with an introduction by A.J.P. Taylor, translated from the German by Samuel Moore, Penguin Books, Harmondsworth.
Maslow, A. (1954) *Motivation and Personality*, Harper & Row, New York.
Matthews, F. (1988) 'Deep Ecology: Where all Things are Connected', *Habitat*, October: 9–12.
Matthews, J. (1994) 'Little World Banks,' in K. Danaher (ed.), *50 Years is Enough: The Case against the World Bank and the International Monetary Fund*, South End Press, Boston.
Mayer, H. (ed.) (1969) *Australian Politics: A Second Reader*, Cheshire, Melbourne.
McCright, A.M. and Dunlap, R.E. (2003) 'Defeating Kyoto: The Conservative Movement's Impact on U.S. Climate Change Policy', *Social Problems*, 50, 3: 348.

McCulloch, J. (1997) 'Mining Asbestos in South Africa', in G. Peters and B. Peters (eds), *The Treatment and Prevention of Asbestos Diseases*, Vol. 15 of the *Sourcebook of Abestos Diseases*, Lexis Law Publishing, Charlottesville.

McEachern, D. (1980) *A Class Against Itself: Power and the Nationalisation of the British Steel Industry*, Cambridge University Press, Cambridge, Chapter 2.

—— (1991) *Business Mates: The Power and Politics of the Hawke Era*, Prentice-Hall, Sydney.

—— (1993) 'Environmental Policy in Australia 1981–1991: A Form of Corporatism?', *Australian Journal of Public Administration*, 52, 2: 173–86.

—— (1995) 'Mining Meaning from the Rhetoric of Nature: Australian Mining Companies and Their Attitudes to the Environment at Home and Abroad', *Policy Organisation and Society*, (winter) 10: 48–69.

—— (1997) 'Foucault, Governmentality and the "New" South Africa', in P. Ahluwalia and P. Nursey-Bray (eds), *Postcolonialism: Culture and Identity in Africa*, Nova Publishers, New York.

McSpadden, L. (2002), 'Environmental law and litigation', in J. Barry and E.G. Frankland (eds), *International Encyclopedia of Environmental Politics*, Routledge, London and New York: 174–75.

Meadows, D.H. *et al.* (1972) *The Limits to Growth: A Report for the Club of Rome's Project on the Predicament of Mankind*, Potomac Associates, London.

Mellor, M. (1994) 'Book review of: *The Politics of the Environment* by Stephen C. Young', *Environmental Politics*, fall, 3: 536–7.

Merchant, C. (1992) *Radical Ecology: The Search for a Livable World*, Routledge, New York and London.

—— (ed.) (1994) *Ecology: Key Concepts in Critical Theory*, Humanities Press, New Jersey.

Mies, M. and Shiva, V. (1993) *Ecofeminism*, Fernwood Publications, Nova Scotia.

Miljöpartiert de gröna/Swedish Green Party (2006) 'Party Programme 2006', *Miljöpartiert de gröna* website, http://mp.se/templates/Mct_78.aspx?avdnr=12131&number=34325& avdelning=12146, accessed 30/08/06.

Miller, P. and Rose, N. (1993) 'Governing Economic Life', in M. Gane and T. Johnson (eds), *Foucault's New Domains*, Routledge, London: 75–105.

Mojavu, M. (2003) *In Nigeria the Flares of Shell are Flames of Hell*, http://www.zmag.org/ sustainers/content/2003-08/10majavu.cfm, accessed 03/09/2006.

Moore, P. (2005) 'Patrick Moore Endorses Nuclear Energy before US Congress', 28 April, 'Statement to Congressional Subcommittee on Nuclear Energy', before the United States Senate Committee on Energy and Natural Resources, http://www.greenspirit.com/logbook. cfm, accessed 1/05/06.

Morris, A. and Herring, C. (1987) 'Theory and Research in Social Movements: A Critical Review', *Annual Review of Political Science*, 2: 137–98.

Moyo, P., O'Keefe and Middleton, N. (1993) *The Tears of the Crocodile: From Rio to Reality in the Developing World*, Pluto Press, London.

Mulligan, D. (ed.) (1996) *Environmental Management in the Australian Minerals and Energy Industries: Principles and Practices*, UNSW Press, Sydney.

Myers, N. (1995) *Environmental Exodus. An Emergent Crises in the Global Arena*, Climate Institute, United States.

Naanen, B. (1995) 'Oil Producing Minorities and the Restructuring of Nigerian Federation: The Case of the Ogoni People', *Journal of Commonwealth and Comparative Studies*, 33, 1.

Naess, A. and Rothenberg, D. (1989) *Ecology, Community and Lifestyle*, Cambridge University Press, Cambridge.

National Wildlife Federation (1993) 'NAFTA and the Environmental Side Agreements: Statement of Dr Jay D. Hair, President and CEO National Wildlife Federation, Washington DC', press release (14 September).

Neidhardt, F. and Rucht, D. (1990) 'The Analysis of Social Movements: The State of the Art and Some Perspectives for Further Research', in D. Rucht, *Research in Social Movements: The State of the Art*, Westview Press Frankfurt, Campus/Boulder, CO.

New Internationalist (2006), 391, July.

New Perspectives Quarterly (1996) 'The Greening of NATO', *New Perspectives Quarterly*, 13, 1: 52–3.

Newman, P. and Kenworthy, J. (1992) *Winning Back the Cities*, Australian Consumers Association and Pluto Press, Sydney.

Nogueira, A. (2002) 'The Birth and Promise of the Indymedia Revolution', in B. Shepard and R. Hayduk (eds), *From ACT UP to the WTO: Urban Protest and Community Building in the Era of Globalisation*, Verso, London and New York: 290–7.

North, R. (1995) *Living on a Modern Planet: A Manifesto for Progress*, Manchester University Press, Manchester.

O'Brien, R. and Williams, M. (2004) 'Governing the Global Political Economy', Chapter 11 of *Dynamics*, Palgrave Macmillan, New York: 287–314.

O'Connor, K. and Sabato, L.J. (1995) *American Government: Roots and Reform* (2nd edn), Allyn & Bacon, Boston, MA.

Oberschall, A. (1993) *Social Movements: Ideologies, Interests and Identities*, Transaction Publishers, New Brunswick and London.

Osaghae, E.E. (1996) 'Human Rights and Ethnic Conflict Management', *Journal of Peace Research*, 33, 2: 171.

—— (1998) *Crippled Giant: Nigeria Since Independence*, Indiana University Press, Bloomington.

Oskwa, B. (2005) 'The Ogoni Ten Years on', pamphlet produced by ERA, Abija, Nigeria.

Overton, P. (2006) 'The Nuclear Solution', Channel Nine, *60 Minutes*, Australian edition, producers Hamish Thomson and Julia Timms, 30 April.

Paehlke, R. and Torgerson, D. (eds) (1990) *Managing Leviathan: Environmental Politics and the Administrative State*, Belhaven Press, London.
—— (eds) (2005) *Managing Leviathan: Environmental Politics and the Administrative State*, 2nd edn, Broadview Press, Peterborough, Canada.
Paehlke, R. and Vaillancourt Rosenau, P. (1993) 'Environment/Equity: Tensions in North American Politics', *Policy Studies Journal*, 21, 4: 672–86.
Paige, S. (1998) 'The "Greening" of Government (Environmental Movement Gains Power in the US)', *Insight on the News*.
Pakulski, J. (1991) *Social Movements: The Politics of Moral Protest*, Longman Cheshire, Melbourne.
Papadakis, E. (1993) *Politics and the Environment: The Australian Experience*, Allen & Unwin, Sydney.
Parsons, T. (1957) 'The Distribution of Power in American Society', *World Politics*, 10: 123–43.
Pearce, D. (ed.) (1991), *Blueprint 2: Greening the World Economy*, Earthscan Publications, London.
Pepper, D. (1984) *The Roots of Modern Environmentalism*, Croom Helm, London.
—— (1993) *Eco-Socialism: From Deep Ecology to Social Justice*, Routledge, London and New York.
Pittock, A.B. (2005) *Climate Change: Turning Up the Heat*, Earthscan and CSIRO Publishing, London and Melbourne.
Piven, F.F. and Cloward, R.A. (1979) *Poor People's Movements: Why They Succeed, How They Fail*, Vintage Books, New York.
Plamenatz, J. (1973) *Democracy and Illusion*, Longman, London.
Player, I. (1997) *Zululand Wilderness: Shadow and Soul*, David Philip, Claremont.
Plumwood, V. (1988) 'Women, Humanity and Nature', *Radical Philosophy*, 48: 6–24.
Poguntke, T. (1992) 'Unconventional Participation in Party Politics: The Experience of the German Greens', *Political Studies*, 15: 239–54.
Polsby, N.W. (1963) *Community Power and Political Theory*, Yale University Press, New Haven, London.
Porritt, J. (1984) *Seeing Green: The Politics of Ecology Explained*, Basil Blackwell, New York and Oxford.
Porter, G. and Welsh Brown, J. (1991) *Global Environmental Politics: Dilemmas in World Politics*, Westview Press, Boulder.
Post Carbon Institute (2006) 'What Is Peak Energy?', Post Carbon Institute website, http://www.postcarbon.org/informed/peakenergy, accessed 5/09/06.
Powers, M. (1993) 'Now What? The Environment', *Human Ecology Forum*, 21, 1: 20–3.
Prendiville, B. (1994) *Environmental Politics in France*, Westview Press, Boulder, San Francisco and Oxford.
President's Council on Sustainable Development (1996) *Final Report*, http://www.whitehouse.gov/WH/EOP/pcsd/Council-report, March.

Preston-Whyte, R.A. (1995) 'The Politics of Ecology: Dredge-mining in South Africa', *Environmental Conservation*, 22, 2: 151–6.

Princen, T. and Finger, M. (1994) *Environmental NGOs in World Politics: Linking the Local to the Global*, Routledge, London and New York.

Rawcliffe, P. (1995) 'Making Inroads: Transport Policy and the British Environmental Movement', *Environment*, 37, 3: 16–32.

Renner, M. (1996) 'The Decline of Nations and the Future of the UN', *World Watch*, 9, 2: 2.

Richardson, D. and Rootes, C. (1995) *The Green Challenge: The Development of Green Parties in Europe*, Routledge, London and New York.

Richardson, G. (1994) *Whatever It Takes*, Bantam Books, Sydney.

Rohrschneider, R. (1991) 'Public Opinion Toward Environmental Groups in Western Europe: One Movement or Two?', *Social Science Quarterly*, 72, 2: 251–66.

—— (1993) 'Environmental Belief Systems in Western Europe: A Hierarchical Model of Constraint', *Comparative Political Studies*, 26, 1: 3–29.

Rootes, C. (1995) 'Environmental Consciousness, Institutional Structures and Political Competition in the Formation and Development of Green Parties', in D. Richardson and C. Rootes (eds), *The Green Challenge: The Development of Green Parties in Europe*, Routledge, London and New York.

Rootes, C. (2006) 'Facing South? British Environmental Organisations and the Challenge of Globalisation,' in Doherty and Doyle (eds), pp. 768–86.

Rosen, R. (1994) 'Who Gets Polluted: The Movement for Environmental Justice', *Dissent*, spring: 223–30.

Rosen, S.J. and Nolan, T. (1994) 'Seeking Environmental Justice for Minorities and Poor People', *Trial*, December: 50–5.

Rosenbaum, W. A. (1991) *Environmental Politics and Policy* (2nd edition), Congressional Quarterly, Washington DC.

Routledge, P., Nativel, C. and Cumbers, A. (2006) 'Entangled Logics and Grassroots Imaginaries of Global Justice Networks', *Environmental Politics*, special issue guest edited by B. Doherty and T. Doyle, 'Beyond Borders: Environmental Movements and Transnational Politics', 15, 5: 839–59.

Roy, A. (1999) *The Greater Good*, India Book Distributors, Bombay.

Rudig, W. and Franklin, M. (1992) 'Green Prospects: The Future of Green Parties in Britain, France and Germany', in W. Rudig (ed.), *Green Politics Two*, Edinburgh University Press, Edinburgh: 37–58.

Rumley, D. (2005) 'The Geopolitical of Global and Indian Ocean Oil Pollution', in Rumley, D. and Chaturvedi, S. (eds), *Energy Security and the Indian Ocean Region*, South Asian Publishers, Delhi.

Rumley, D. and Doyle, T. (2005) 'The Uranium Trade in the Indian Ocean', paper presented at the Third Indian Ocean Research Group (IORG) meeting, Kuala Lumpur, Malaysia, July.

SAFF Australian Farmers' Fuel (2006) 'What Is Bioethanol?', SAFF website, http://www. farmersfuel.com.au/, accessed 5/09/06.

SAFF sponsor representative (2006), presentation at 'Peak Oil or Oil Shock?', lecture, UniSA Hawke Centre, 28 August.

Sale, K. (1993) *The Green Revolution: The American Environment Movement 1962–92*, Hill & Wang, New York.

Salleh, A. (1992) 'The Ecofeminism/Deep Ecology Debate: A Reply to Patriarchal Reason', *Environmental Ethics*, fall: 195–216.

Sandbrook, R. (1993) 'Live and Learn', *New Statesman and Society*, January: 29–30.

Sandweiss, S. (1998) 'The Social Construction of Environmental Justice', in D.E. Camacho (ed.), *Environmental Injustices, Political Struggles: Race, Class, and the Environment*, Duke University Press, Durham and London: 31–57.

Scarce, R. (1990) *Eco-Warriors: Understanding the Radical Environmental Movement*, Noble Press, Chicago.

Schlosberg, D. (2004) 'Reconceiving Environmental Justice: Global Movements and Political Theories, *Environmental Politics*, 13, 3: 517–40.

Schmid, J. (1999) 'Pacifists Speaking Up in Bonn: Bombing Reported Causing "Unrest" in Greens Party', *International Herald Tribune*, 1 April, p. 2.

Schrader, R. (1992) 'The Fiasco at Rio', *Dissent*, fall: 431.

Schrecker, Ted (1990) 'Resisting Environmental Regulation: The Cryptic Pattern of Business-Government Relations', in R. Paehlke and D. Torgerson (eds), *Managing Leviathan: Environmental Politics and the Administrative State*, Belhaven Press, London: 165–99.

Schreiber, H. (1995) 'The Threat for Environmental Destruction in Eastern Europe', *Journal of International Affairs*: 361–91.

Seliger, M. (1976) *Ideology and Politics*, Allen & Unwin, London.

Sessions, G. and Naess, A. (1983) 'The Basic Principles of Deep Ecology', *Earth First*.

Shiva, V. (1991) *Ecology and the Politics of Survival: Conflicts Over Natural Resources in India*, United Nations University Press, Tokyo, http://www.unu.edu/unupress/unupbooks/ 80a03e/80A03E00.htm, accessed 8 May 2006.

—— (1994) 'Development, Ecology and Women', in C. Merchant (ed.), *Ecology: Key Concepts in Critical Theory*, Humanities Press, New Jersey.

—— (ed.) (1994) *Close to Home: Women Reconnect Ecology, Health and Development Worldwide*, New Society Publications, Philadelphia.

—— (2006) 'Monsanto's Genetic Engineering Trials in India are Dangerous and Anti-democratic', *Navdanya* website, http://www.navdanya.org/articles/ge_trials.htm

Sierra Club (1997) '"Local Control" A Smokescreen for Logging', *The Planet: The Sierra Club Activist Resource*, http://www.sierraclub.org/planet/ 199711/delbert.html

—— (2006a) 'Clean Air: Clear Skies Proposal Weakens the Clean Air Act', Sierra Club website, http://www.sierraclub.org/cleanair/clear_skies.asp, accessed 19/09/06.

—— (2006b) 'Forest Protection and Restoration: Debunking the "Healthy Forests Initiative"', Sierra Club website, http://www.sierraclub.org/forests/fires/healthyforests_initiative.asp, accessed 19/09/06.

Sierra Defense Legal Fund (2005) 'Environmental Groups Launch Federal Court Case in Historic Bid to Save Spotted Owl: BC Government's Refusal to Protect Owl Results in Landmark Legal Challenge', 6 December, http://www.sierralegal.org/m_ archive/pr05_ 12_06.html, accessed 10/09/06.

Simon, J. and Kahn, H. (1984) *The Resourceful Earth*, Basil Blackwell, Oxford.

Simpson, A. (1999) 'Buddhist Responses to Globalisation: Thailand and Ecology – The Yadana Gas Pipeline', in S. Sivaraksa (ed.), *Socially Engaged Buddhism for the New Millennium*, The Sathirakoses-Nagapradipa Foundation and Foundation for Children, Bangkok.

—— (2006) 'Downfall? Capitalism and Dissent in Thaksin's Thailand', Refereed paper presented at the Second Oceanic Conference on International Studies. 5–7 July, Melbourne: University of Melbourne, http://www.politics.unimelb.edu.au/ocis/, accessed 7 March.

—— (2007) 'Gas Pipelines and Environmental Security in South and Southeast Asia: Who Wins and Who Loses?', in T. Doyle and M. Risely (eds), *Crucible for Survival: Environmental Security in the Indian Ocean Region*, Rutgers University Press, New Brunswick, New Jersey and London, forthcoming.

Singh, M. (1996) 'Environmental Security and Displaced People in Southern Africa', *Social Justice*, 23, 4: 125–9

Skene, C. (2003) 'Authoritarian Practices in New Democracies', *Journal of Contemporary Asia*, 33, 2: 189.

Skinner, J. (1995) 'Green Party Revival', *The Progressive*, 59, 5: 14.

Smith, G. (2002) 'Agenda 21', J. Barry and E.G. Frankland (eds), *International Encyclopedia of Environmental Politics*, Routledge, London and New York, pp. 5–6.

Snow, D. (ed.) (1992a) *Inside the Environmental Movement: Meeting the Leadership Challenge*, Island Press, Washington, DC, and Covelo, CA.

—— (ed.) (1992b) *Voices from the Environmental Movement: Perspectives for a New Era*, Island Press, Washington, DC, and Covelo, CA.

St Clair, J. (2004) 'The Demise of the Green Party', *Synthesis/Regeneration 35: A Magazine of Green Social Thought*, fall: 30–1.

Strong, G. (1998) 'The Green Game', *The Age*, 17 August, http://www.theage.com.au/daily/ 980717/news/news22.html

Subcomandante Insurgente Marcos (1997), 'On Independent Media', in Ponce de León, J. (ed.) (2002), *Our Word is Our Weapon: selected writings*, Seven Stories Press, New York: Chapter 36.

Sykes, T. (1999) 'How activists have hijacked the annual meeting', *Australian Financial Review*, 11 July.

Sylvan, R. and Bennett, D.H. (1986) 'Deep Ecology and Green Politics', *The Deep Ecologist*, 20, April.

Szabo, M. (1994) 'Greens, Cabbies and Anti-Communists: Collective Action During the Regime Transition in Hungary', in E. Larana *et al.* (eds), *New Social Movements: From Ideology to Identity*, Temple University Press, Philadelphia.

Tang, J. (2004) 'Brief Analysis of Public Participation in China', paper presented in School of Social Sciences, Discipline of Geography and Environmental Studies, University of Adelaide.

Tao, Fu, (2005) 'Respite for Yunnan Anti-Dam NGO as Movement Mourns Loss of Key Activist,' *China Development Brief*, January, http://www.chinadevelopmentbrief.com/ node/104, accessed 16 August 2006.

Target Earth (n.d.) 'Serving the Earth, Serving the Poor', http://www.targetearth.org/

Taylor, B. (1991) 'The Religion and Politics of Earth First!', *The Ecologist*, 21, 6: 258–66.

The 3rd Citizen's Conference on Dixon and Other Synthetic Hormone Disruptors (1996) Time for Action, Baton Rouge, LA.

The Weekend Australian (2006) 14–15 January.

Wilderness Society (2004a) 'Green Groups Cautiously Welcome Latham's Plan', media release, TWS website, http://www.wilderness.org.au/campaigns/policy/elections/federal/ 2004/alp051004/, accessed 5/10/04.

—— (2004b) 'Undecided Voters Urged to Vote for Tasmania's Forests', media release, TWS website, http://www.wilderness.org.au/campaigns/policy/elections/federal/2004/voters/, accessed 8/10/04.

—— (2004c) 'Tasmanians Vote to Save the Forests', media release, TWS website, http:// www.wilderness.org.au/campaigns/policy/elections/federal/2004/tas_vote_2_save_forests/, accessed 12/10/04.

Thomas, I. (1996) *Environmental Impact Assessment in Australia: Theory and Practice*, Federation Press, Sydney.

Thompson, H. (1993) 'Malaysian Forestry Policy in Borneo', *Journal of Contemporary Asia*, 23, 4: 503–14.

Toffler, A. (1970) *Future Shock*, Pan, London.

Torgersen, D. (2006) 'Expanding the Green Public Sphere: Post-colonial Connections', *Environmental Politics*, special issue guest edited by B. Doherty and T. Doyle, 'Beyond Borders: Environmental Movements and Transnational Politics', 15, 5: 713–30.

Touraine, A. (1981) *The Voice and the Eye: An Analysis of Social Movements*, Cambridge, Cambridge University Press.

United Nations Conference on Environment and Development (1992) Agenda 21, Geneva (on line at http://www.erin.gov.au/portfolio/esd/nesd/Agenda21/html).

Wald, M. (2006) 'Ex-Environmental Leaders Tout Nuclear Energy', *The New York Times*, 25 April.

Wall, D. (1999) *Earth First! and the Anti-Roads Movement: Radical Environmentalism and Comparative Social Movements*, Routledge, London and New York.

Ward, B. and Dubos, R. (1972) *Only One Earth*, W.W. Norton, New York.
Ward, C. (2004) *Anarchism: A very Short Introduction*, Oxford University Press, New York.
Weatherly, C. (1993) 'The Ecological Imperative and Political Success: A Paper for *Ecopolitics VII*', Abstract for Ecopolitics VII Conference, 2 July.
Weber, M. (1978) *Economy and Society: An Outline of Interpretive Sociology*, edited by G. Roth and C. Wittich, University of California Press, Berkeley.
Wilson, E.O. (1992), *The Diversity of Life*, Belknap Press, Harvard.
Women's Environment and Development Organization (WEDO) (1995) 'Transnational Corporations at the UN: Using or Abusing their Access?', *Newsletter*, 2: 2–4.
World Commission on Environment and Development (WCED) (1990) Chaired by Gro Harlem Brundtland and the Commission for the Future, *Our Common Future* (Australian edition), Oxford University Press, Melbourne.
Young, J. (1990) *Post Environmentalism*, Belhaven Press, London.
Yunus, M. (1999) 'The Grameen Bank', *Scientific American*, November: 91–5.

Index